# To So Few

## The Trial

# Books by Cap Parlier:

—

**<u>Anod series</u>**
**The Phoenix Seduction** (1995)
**Anod's Seduction** (2004) [reprint of The Phoenix Seduction]
**Anod's Redemption** (2004)

—

**Sacrifice** (2000)
**The Clarity of Hindsight** (2016)

—

**<u>To So Few series</u>**
**To So Few – In the Beginning** (2014)
**To So Few – The Prelude** (2014)
**To So Few – Explosion** (2015)
**To So Few – The Trial** (2016)

—

<u>and with **Kevin E. Ready:**</u>
**TWA 800 - Accident or Incident?** (1998)

—

Coming soon from Cap Parlier, **To So Few – The Verdict**,
the fifth book of the series novel of flight and a warrior's life.

—

# To So Few

## The Trial

### by

# Cap Parlier

**SAINT GAUDENS PRESS**
Wichita, Kansas  & Santa Barbara, California

**Saint Gaudens Press**
Post Office Box 405
Solvang, CA 93464-0405

Http://www.SaintGaudensPress.com

Saint Gaudens, Saint Gaudens Press
and the Winged Liberty colophon
are trademarks of Saint Gaudens Press



Print edition ISBN: 978-0-943039-33-6

Library of Congress Catalog Number - 2016914883

Printed in the United States of America

The TO SO FEW series books are works of fiction.  Any reference to real people, objects, events, organizations, or locales is intended only to give the fiction a sense of reality and authenticity.  Other names, characters and incidents are the products of the author's imagination and bear no relationship to past events, or persons living or deceased.

# Dedication

—

To all those who have gone before us, and given their last full measure of devotion to the cause of freedom and the defense of our liberty.

# Acknowledgments

—

To John Richard and Roger Benefiel for research assistance.

To my wife, spouse, partner, sponsor and cheerleader, Jeanne, who tolerated the hours, days, weeks, months and years of research, writing and discussion.  She has and continues to tolerate my love of flight and the need to tell a story about the greatest event in human flight.

To my reviewers: my wife, Jeanne; John Richard; and Leta Buresh, for their patience, reflection, opinions and suggestions.  I believe they made it a stronger story.

A special recognition must be offered to numerous individuals who provided their knowledge, experience and precious time to assist my historical research.
-- Imperial War Museum – Dr. Neil Young, Research and Information Office
-- Royal Air Force Museum – Mr. Mungo Chapman, Research & Information Services
-- Duxford Aerodrome Museum
-- R.J. Mitchell Memorial Museum – Mr. D.G. Upward, Director
-- National Railway Museum – Mr. Philip Atkins, BSc, Librarian
-- Churchill Archives Centre – Ms. Carolyn Lye
-- House of Lords – Mr. D.L. Prior, Record Office

If there are errors in the representation of historical details, the responsibility rests solely with me and must in no way reflect upon the experts acknowledged above.

# Prologue

—

Since the opening shots of war in Europe and the Allied declaration of war on Germany two days later, the large British and French air and land forces remained in defensive position in Northeast France, as the German *Wehrmacht* (Armed Forces) rampaged through Poland and consolidated their conquest to the River Bug. By their prior non-aggression pact and in concert with the Germans, the Soviet Union invaded and occupied Eastern Poland and the Baltic countries. The allies did not move to aid the Poles, as if they had been shocked and frozen by the audacity and very rapid success of the new, armored, mobile warfare the Press dubbed *Blitzkrieg* (Lightning War). After the month of German and Soviet conquest in the east, the allies still could not bring themselves to move against the Germans for the next six (6) months in a period the Press sarcastically called the Phony War or *Sitzkrieg*. The lull allowed the Germans to refit, rearm, refresh and redeploy the bulk of their armed forces, and prepare for their planned spring offensive to the north and west.

The Soviets encouraged by their autumn gains sent the Red Army into neighboring Finland during the winter months and quickly learned the grossly outnumbered Finns would not be so easy to defeat. After terrible losses, the Red Army withdrew and a tenuous peace treaty was signed at the beginning of spring.

In the early dawn hours of Friday, 9.April.1940, without provocation, the Germans invaded neighboring Denmark and Norway. The Danes were overwhelmed in a day. It would take three months to defeat the Norwegians. The Germans promptly installed a well-known Norwegian fascist collaborator as titular head of the occupation government and added a dedicated Nazi administrator along with a *SchutzStaffel* (SS) general as the fist to maintain order. The British deployed modest Navy, Army and Air Force units in a meager effort to sustain the Norwegians; but their largely symbolic endeavor paled against the German force they faced.

Then, at dawn on Friday, 10.May.1940, a major, combined arms, German force invaded neighboring Holland and Belgium. This time, the Allies moved. The British Expeditionary Force and French *le Armée de Terre* advanced into Belgium to stop the German attack and defend their allies. Three days later, with those forces engaged, the German shocked the world as a massive armor phalanx burst out of the Ardennes Forest into the gap between the impregnable Maginot Line defensive fortifications and the advancing Allied

Northern Force.  Despite the advance notice of the Polish campaign, the Allies could not find the means to stop the German tanks from reaching the English Channel coast and cutting off the entire Northern Force from reinforcement or resupply.

As the German forces poured across the western frontier that Friday in May, King George VI reluctantly asked The Right Honorable and Gallant Winston Leonard Spencer Churchill, CH, Member of Parliament for Epping, to form a wartime coalition unity cabinet and lead His Majesty's Government in the defense of the United Kingdom and the greater empire beyond the Home Islands.  Churchill acted quickly, decisively and masterfully to form a core, leadership, War Cabinet.  In addition to being the King's first minister and leader of the coalition War Cabinet, Prime Minister Churchill became the Leader of the Conservative Party, Leader of Commons, and appointed himself 1$^{st}$ Lord of the Treasury and the newly created position of Minister of Defense over the other historical military department ministers.  He intended to take a very intimate and personal role in waging war successfully.  From a War Cabinet leadership perspective alone, Churchill faced a daunting challenge beyond the warfighting, swiftly closing in on Great Britain.

The situation in France deteriorated rapidly in May 1940.  Several attempts by both the British and French forces on both flanks of the armored penetration failed to stop the Germans from cutting off and isolating the British Expeditionary Force.  Surrounded and with the collapse of France rushing toward them, Prime Minister Churchill made the momentous decision to evacuate the British Expeditionary Force in what became known as the Miracle of Dunkirk – a third of a million British and Allied soldiers were plucked from the northeastern beaches to fight another day.  The British also stoically prepared and waited for the German invasion that was sure to come.

The staggering events in Europe produced more than a little concern across the Atlantic Ocean.  President Franklin Delano Roosevelt had his own struggles as he sought to end the Great Depression in the United States; yet, he recognized the reality of the naked aggression and oppression of the Germans now occupying most of the European Continent along with their Italian ally.  Roosevelt masterfully walked a very thin line between the very strong isolationist forces inside and outside the U.S. Government, and the pleas for support from his friend Winston Churchill, as he used old, obscure laws to bypass the Neutrality Acts and resupply the decimated British Army and Royal Air Force.  He also quietly moved public opinion and legislation to rebuild the woefully ill-prepared U.S. armed forces for what he progressively saw as an inevitable world war the United States would not be able to hide from behind their ocean moats.  President Roosevelt also began making the necessary changes in his Cabinet and staff in preparation for war.

As the immensely successful *Wehrmacht* consolidated their ill-gotten gains on the Continent and prepared for their crossing of the English Channel, the Royal Navy remained dominant on the sea approaches, but found itself playing a desperate catch-up effort as the German Navy submarine service achieved surprising successes in their operations to strangle the British.  Despite the dark days of war, the Royal Navy was not without their successes as they defeated the surface raider *DeutscheKriegsMarine Panzerschiff Admiral Graf Spee* in the Battle of the River Platte (17.December.1939) and the bold neutralization of the Vichy French *Marine Nationale* at Mers-el-Kébir, Algeria, in Operation CATAPULT (3.July.1940).  Yet, despite the dominance of the Royal Navy, the ships needed air superiority to defend the English Channel.  The air cover for the Royal Navy and the air defense of Great Britain fell to the Royal Air Force and specifically Fighter Command.  Yet, the nation's air defense during the foreboding summer of 1940 would ultimately stand upon the skills of nearly 600, young pilots in single-seat fighter airplanes, with an average age of less than 20 years, working together to face an overwhelming *Luftwaffe* in both numbers and combat experience.

On Monday, 10.July.1940, the Germans began their preparatory air offensive against Great Britain by attacking ships transiting the English Channel as much to draw Fighter Command into decisive engagement as to deny Channel waters to Allied merchant vessels and warships.  The Fighter Command pilots learned quickly they faced a formidable adversary.  The Germans soon learned the British would not be quite so easy to subdue.

—

Into this history came two friends and brothers-in-arms Brian Arthur 'Hunter' Drummond and Jonathan Andrew Xavier 'Harness' Kensington.  Brian was a young American volunteer from the Great Plains of Kansas, who defied U.S. federal law to join the Royal Air Force before the war began.  Jonathan came from an upper middle class family near Newcastle-upon-Tyne.  The two pilots met during advanced flight training, became close friends, were commissioned as pilot officers in the Royal Air Force, and fortunately were assigned to the same fighter squadron – No.609 (West Riding of Yorkshire) Auxiliary Squadron – at RAF Drem in December 1939.  When the Germans invaded France and the Low Countries, the squadron moved south from Scotland to be stationed at RAF Middle Wallop, just north of Southampton.

By the end of July, 1940, as the intensity of the aerial battle increased rapidly, just short of a half dozen squadron pilots had lost their lives in aerial combat: Flying Officer George 'Angle' Ashcroft; Pilot Officers Reginald 'Organ' Foxworth and Stephen 'Mongo' Strickland; Flying Sergeant James 'Junior' Carrolton; and one of the American volunteer aviators Pilot Officer Henry 'Hank' Maxwell, who lasted just two weeks after joining the squadron. Replacement pilots began arriving in England to help make good the losses suffered by No.609 Squadron and the rest of Fighter Command.  The surviving No.609 Squadron members and their flight positions were:

<u>'A' Flight, Blue Section:</u>
Squadron Leader Horatio Michael 'Spike' Darling (English)
Pilot Officer Roland 'Boxer' Stockard (English)
Pilot Officer Kormer 'Curly' Mansek (Czechoslovak)
<u>'A' Flight, Green Section:</u>
Flight Lieutenant Roger 'Jackstay' Beamish (Scottish)
Pilot Officer Janus 'Crazy' Kradilcek (Czechoslovak)
Pilot Officer Drummond (American)
<u>'B' Flight, Red Section:</u>
Flight Lieutenant Robert Gates 'Sparky' Morrow (English)
Pilot Officer Frank Oscar 'Red' Burns (American)
Pilot Officer Kensington (English)
<u>'B' Flight, Yellow Section:</u>
Flight Lieutenant John 'Waggle' Davies (Welsh)
Flying Sergeant Miles 'Fog' Johnson (English)
Pilot Officer Stanley Jordan 'Slim' Koenig (American)

Flying Officer James Royster served as the squadron's intelligence officer.  Corporal Jennifer Warren, Women's Auxiliary Air Force (WAAF) was the able Dispersal Operations Clerk – the dependable anchor for the squadron's operations.  'Harness' Kensington had been hand selected to represent operational pilots as one of the exploitation pilots evaluating captured and refurbished German aircraft.  'Hunter' Drummond had become an ace with 6½ credited victories in aerial combat, although his accomplishment had gone unnoticed in the fury of battle, and he was now recovering in hospital after having his third aircraft shot out from around him and being miraculously saved by Charlotte Grace Palmer née Tamerlin – a courageous and resourceful war widow, and the farmer-owner into whose pond Brian parachutred into in an unconscious state.

The combat losses for Fighter Command mounted sharply.  Refugee pilots from France made their way to England, some of them staying a step ahead of the Germans from the rape of Czechoslovakia, through the invasion of Poland, and now the Low Countries and France itself.  More than a dozen countries from the corners of the British Empire and many of the occupied countries of Europe now flew with their British brethren, including more than a handful of American volunteers.  Despite the welcome additions, the challenge for the leadership of Fighter Command, the entire Royal Air Force and ultimately for His Majesty's Government came down to managing the net equation for both pilots and aircraft as the Germans used their advantage to bleed the British.

And so, here begins our story.

—

# Chapter 1

If all men were friends,
there would be no need for justice.

-- Aristotle

*Sunday, 4 August 1940*
*RAF Middle Wallop*
*Middle Wallop, Hampshire, England*

### Week 5

**P**ilot Officer Jonathan 'Harness' Kensington began the first Sunday in August in calm. The pilots did not get their wake-up call until 06:30. The sun was well above the horizon, and the scattered clouds made for a fine summer's day. Flight Lieutenant Roger 'Jackstay' Beamish and Pilot Officer Roland 'Boxer' Stockard added their observations that the late wake-up call offered up suspicions about what might be coming their way. They had been awakened at 04:30 to 05:00 virtually every morning since the invasion of Denmark and Norway four months ago.

Everything this summer morning moved in slow motion. People moving through the sector control station facility did not appear to have the usual energy to their step. As the No.609 Squadron pilots arrived at their Dispersal building, they settled into the odd array of chairs randomly assembled outside in the sunlight where they left them the previous evening. Only Squadron Leader Horatio 'Spike' Darling joined Corporal Jennifer Warren, WAAF, inside the humble building.

Half the pilots were asleep when Darling returned to provide a report of status and plans. Routine operations. Nothing exciting or extraordinary to be expected today. A few of the lot perked up when their assigned intelligence officer, Flying Officer James Royster, appeared to repeat the latest situation report from the Air Ministry and Fighter Command. Royster's words did not quite match the pace of the day so far. Although his manner was less than animated, he seemed to be saying they expected more action and a higher possibility of invasion. The words just did not seem to fit the mood of the moment.

"Maybe this is the quiet before the storm," said Flight Lieutenant 'John 'Waggle' Davies with his eyes still shut against the sunlight.

"Or, the eye of the storm," offered 'Red' Burns with his eyes also still closed.

"Either way, it means rough weather ahead, laddies," added 'Jackstay' Beamish.

"Not to worry," said 'Sparky' Morrow, "we have been handling things fairly well."

Jonathan sat up sharply, surprised at the flight lieutenant's words and wanted to remind him of the lost lives, damaged aircraft and injured comrades. His best friend and one of the best fighter pilots in the squadron, the first American volunteer, Brian 'Hunter' Drummond, still lay in a hospital bed having, only by the grace of God, survived what should have been a fatal explosion. Jonathan Kensington wanted to remind him of the facts. With the mood of the day, he held his tongue, and he soon settled back into his floppy beach chair.

"Could be," said Beamish, "but, what about 'Hunter' and the others? It doesn't seem to me we've done that well."

"How is Brian, Jonathan?" asked Stockard.

Several heads rose up just enough to look at Jonathan. "He is all right, I suppose. He still feels the effects of a fairly good clot on the gourd. He says the surgeons will let him go in a day or so."

"Wasn't he still unconscious when you went to see him Thursday morning, Skipper?" asked 'Slim' Koenig.

"Yes, he was."

"That must have been a pretty good bump to leave him out that long."

"Maybe if things stay this slow, we can all go see him," offered Koenig.

"What time do they close the place to visitors?" asked Morrow.

"Oh, hell, Rob, so we show up a little late. What are they going to do, arrest us?" responded Davies.

Nearly everyone laughed at the thought of the entire fighter squadron's pilots being arrested for trying to see their injured comrade during the Battle of Britain. Not a likely scenario, so most joked.

"Perhaps we can avoid the possibility of strutting our stuff if we are released early," Beamish said.

The telephone rang stopping the jovial conversation instantly. "Squadron Leader Darling, it is for you, sir," announced Corporal Warren from her desk inside Dispersal.

Darling disappeared for a quarter of an hour and the pilots remained silent as they waited for whatever news there might be. None of the other Middle Wallop squadrons had taken to the air yet this morning which with such a fine morning was rather unusual. The only sounds of flight had been the last few initial engine run-ups to prepare the fighters for the day's operations. Only the chirping of various birds filled the air when Squadron Leader Darling returned to the group.

"Well, gentlemen, for some odd reason, Gerry seems to have taken a holiday. Several small convoys transit the Channel today. Routine patrols, I'm afraid."

The grumbles and grunts marked objections to the disturbance rather than any real or perceived danger. Patrols were boring. While flying was indeed flying, and that was generally good – patrols were still boring. Squadron Leader Darling completed a leisurely briefing that established the size, location, course and speed for each of the convoys in or approaching the English Channel. The Middle Wallop Sector squadrons would have responsibility for Convoy AGENT that consisted of sixteen merchants and two destroyers moving west at eight knots and would be just east of Portsmouth at mid-day. This particular convoy was slow by convoy norms and had already been attacked several times during their transit from the River Thames Estuary and through the Dover Straight westbound.

"There you have it, lads. 'B' Flight is expected overhead at half eleven. Remember to throttle back, 'Sparky.' We must stretch the petrol today. Two hours on station. 'A' Flight to relieve at half one. Initial patrol altitude should be angels two oh. Any questions?"

No one spoke up. With another hour to launch time to make their overhead time, everyone settled back into their chairs. Flight Lieutenant Morrow would rouse them at the appropriate time. Squadron Leader Darling left the group walking toward the airfield operations building and tower at the apex of the V of hangars. Jonathan wondered only briefly what business their leader had at Operations, and then allowed the warmth of the Sun and underlying tiredness to carry him into slumber. In what seemed to be only several instants, a kick to the edge of his boot brought Jonathan back to reality.

Flight Lieutenant Robert 'Sparky' Morrow, leader of 'B' Flight – Blue and Yellow Sections – nodded toward the flight line. The silent call to launch like an earlier morning, pre-dawn, wake-up call struck Jonathan as rather odd with the Sun directly overhead. The five other pilots followed their flight leader to the aircraft.

The take-off produced the most excitement for the flight. Feeling the full power of the Merlin III engine along with the hydraulic rush associated with raising the landing gear just after breaking ground and skimming across the grass up to the far treeline pumped in the usual dose of aviator's narcotic. From that point on, everything proceeded at a much slower pace. The throttle came back halfway and the propeller pitch retarded to 2,000 rpm to conserve fuel. The cruise settings produced a slow, monotonous climb until they reached 20,000 feet just before they arrived overhead the bevy of ships on the blue-green water among the clouds below.

Flight Lieutenant Robert 'Sparky' Morrow took the loose combat formation of six Spitfire fighters into a long racetrack pattern orbit. The sky was empty and the radio was unusually quiet.  Sector Control reported several times the lack of any RDF contacts.  No business today.  The low droning of the engine, lack of the whistle of high speed air passing over the canopy and the gentle, shallow turns in their patrol pattern made the constant search of the skies around them very difficult.  Jonathan wanted to go back to sleep.  They probably all did.  Jonathan knew none of them really wanted aerial combat although the pronouncements of the braggart indicated otherwise.  He loved flying as most of the others did, but this uneventful, repetitive, slow mission did not seem like flying to him.

'A' Flight joined them in position before 'B' Flight returned to RAF Middle Wallop.  The first sortie ended with less excitement than it had begun.  The intelligence debriefings were very short.  No action.  The routine continued through two full cycles without seeing or hearing about an enemy aircraft.  As they prepared for the beginning of their third cycle, Jonathan thought about the expenditure of precious fuel.  Was this the best use of their resources?  What was Sector Control worried about?  He could not see the purpose of the constant protective cover when they were fairly good at responding quickly as soon as a German raid was spotted on the Chain Home RDF scopes as a possible threat.  Maybe there was an extra special cargo in the convoy?  Jonathan smiled, that had to be it.

"Here we go, mates," said Morrow, as he walked toward the line of fighters.

Jonathan heard the telephone ring behind them.  Instinctively, he turned around as several of the others did as well.

Corporal Jennifer Warren came running out of the Dispersal building.  "Mister Morrow, Sector has scrubbed the patrol."  She slowed to a walk when she saw everyone stop.  "We are to stand at Available until 'A' Flight returns."

"Hallelujah," proclaimed Flying Sergeant Miles 'Fog' Johnson.  "Those damnable circles were making me dizzy."

They all laughed as they returned to their chairs.  'Slim' Koenig opted for a lush patch of grass as he lay flat and stretched to relax himself.  No one seemed to be particularly talkative as they waited.  The tones of low powered Merlins announced the return of 'A' Flight followed promptly by the popping protests of the idling engines as they landed.  The distinctive red, gunport patches were all still in place.  They had more of the same.

The others joined their comrades in the chairs or grass as Squadron Leader Darling passed through the group going directly to his office.  Several minutes passed before they saw their leader again.  Those that were awake looked at him until they realized he had nothing to say.

"I wonder what the bloody friggin' Nazis are up to?" asked 'Waggle' Davies with his eyes shut and his recumbent body motionless in the lawn chair.

"No good, you can bet," responded 'Boxer' Stockard.

"Probably savin' their juice," added 'Jackstay' Beamish.

"For a go at an invasion."

"That's it," said Davies. "They are saving their energies to try what Napoleon could not do."

"Ought to know better," Stockard said. "You would think the dumb bastard would study his history lesson, wouldn't you."

"Probably thinks he's above it," 'Red' Burns said.

"He is about to get himself bloodied after we give him a clot in the nose," barked Davies.

The laughter punctuated and terminated the brief exchange as they each drifted back into their thoughts. Darling came and went several times without words, like a caged tiger waiting for his keeper to open the gate. The Sun descended behind a row of gathering clouds to the West casting a cool shroud over the countryside. The fourth time ended their duty day as the squadron was told to stand down.

As they hung their flying equipment on the appropriate wall pegs, Jonathan spoke. "Who wants to go see Brian?"

"Can we wait until after meal?" asked 'Crazy' Kradilcek.

"Sure."

Most of the pilots had other commitments for the evening hours. Squadron Leader 'Spike' Darling asked Jonathan to pass along his best wishes. 'Jackstay' Beamish, 'Crazy' Kradilcek and 'Fog' Johnson all agreed to accompany Jonathan down to Hampshire Central Hospital in Winchester. After the evening meal, the automobile journey took half an hour.

The nurses told them they had thirty minutes remaining of visiting hours, which they accepted. The medical staff also told them Brian's recovery continued to progress rapidly, and he should be released for duty tomorrow. As they entered the ward, the smell of disinfectant and decay confirmed the purpose of the place.

Jonathan looked left where he remembered Brian's bed to be, however it was used but empty. They found Brian at the far end of the long rectangular room, talking to one of the other patients. Probably another injured pilot if Jonathan was to guess.

"Brian," shouted 'Jackstay' Beamish, "what are you doing standing around while the rest of us are out defending the faithful?"

"Hey, guys."

"Don't let him fool you, 'Hunter.' We had another set of boring patrols, and no Gerries to shoot at," said 'Fog' Johnson.

They all shook hands, and then walked out of the ward to a lounge area so they would not disturb the other patients anymore.  Roger Beamish passed on the best wishes from their leaders and the others who could not make the visit. The central topic was Brian's health, followed closely by the story of his downing and rescue.

"When 'Harness' and I passed over that pond, a young lady was in there saving your humble arse," Beamish said.

"Yeah, I know.  I met her yesterday, after you left Jonathan.  An incredible woman, if my guess is right."

Several hoots and hollers acknowledged Brian's pronouncement.  The pilots derived great pleasure from accentuating certain key activities that seemed to pervade their experience.  Flying mistakes, close calls, poor judgment and bad luck received the most attention from their brethren.  Outsiders rarely understood the often comical, devil-may-care attitude the pilots had toward their near-death experiences in flight.  The other principal target of their belittlement, humor and gibes was women.  They simply enjoyed bringing females and any relationship into the same arena as flying.  Likewise, few people could appreciate the significance.

"So, tell us about her?" asked Janus Kradilcek.  "We want to know all."

"Not on your life.  I'm not giving you blokes more ammunition."

"Ah, ha.  I do believe we have ourselves a challenge," Beamish said.  "We saw what she did, laddie.  The woman deserves a medal for her selfless effort to save your young arse."

"I know.  She won't talk much about it.  Jonathan has already told me his part.  I suppose I'm lucky to be alive."

"Nonsense, laddie.  Divine providence.  You are destined to greatness in His Majesty's service."

They laughed and joked until the nurse supervisor came to throw them out of the ward.  Fortunately for all of them, Brian's headaches were nearly gone.  The hospital expected to discharge Pilot Officer Brian Drummond tomorrow around mid-day.  Brian's recovery from such a horrific event buoyed all of them.  They stopped at nearly every pub en route back to RAF Middle Wallop to celebrate the pending return of one of their own.  They were all lucky.

———

*Monday, 5 August 1940*
*RAF Middle Wallop*
*Middle Wallop, Hampshire, England*

The obscuration of the summer haze and an occasional small puffball of a cloud were the only detractors from an otherwise warm, not quite hot, humid day. The formalities of Brian Drummond's medical release examination agitated his impatience, but confirmed his readiness to return to the squadron. The driver of the Royal Air Force automobile sent to retrieve him had to wait more than an hour, which did not seem to bother the young man. The journey across Hampshire to join the squadron inched along at the same pace as the medical evaluation. Brian wanted to get back into the air as soon as he could.

As they approached Middle Wallop, a section of three Hurricanes took off low over the trees. The Merlins at full power sounded like the most glorious music to Brian. The itch to feel the power and speed, to smell gasoline fumes as the Merlin coughed to life, and to hear the rush of air past the canopy became unbearable. He hoped Squadron Leader Darling would allow him to fly as soon as he returned to the squadron. The headaches were nearly gone, but he felt ready to fly.

Several of the guards at the main gate recognized Pilot Officer Brian Drummond and welcomed him back. The driver followed Brian's directions precisely and dropped him off at the No.609 (West Riding of Yorkshire) Squadron Dispersal building. Most of the 'PR' Spitfires sat idle on the flight line. Brian did a quick inventory. Flight Lieutenant John 'Waggle' Davies' Yellow Section was missing and probably on patrol.

"The prodigal son returns," Morrow said loudly, as Brian turned the corner to find most of the pilots outside in the sun.

All the pilots stood, shook Brian's hand and began asking him an endless set of unanswered questions. Darling joined the group to add his welcome to Brian. The sensations and feelings of family overwhelmed him. These men had become his family. Then came the fundamental question.

"Are you ready to fly?" asked Squadron Leader Darling.

"You bet."

"No. I mean did the surgeons return you to duty status, and are you physically recovered and ready to fly?"

"Yes, sir. They returned me to duty this morning, and I'm ready right now."

"What say you, 'Jackstay?' Why don't you take up your section and give young 'Hunter' a good wring out?"

"As you say, Skipper. We shall give our young buck a good go."

Pilot Officer Janus 'Crazy' Kradilcek picked up his flying equipment along with Beamish. Corporal Warren had drawn new items for Brian, which he thanked her for doing. He quickly adjusted his leather helmet, flying goggles and oxygen mask to fit his head. Flight Lieutenant Beamish asked Corporal Warren to call in their training mission to Sector Control so they would expect them. The three Green Section pilots walked toward their poised fighters. A brand new Spitfire Mark IA fighter – the fourth aircraft marked with his 'PR-F' letters – waited for Brian along with his ground crew – Leading Aircraftman Bernard 'Bernie' Gordon (crew chief), and Aircraftmen Jordan Toldson (rigger) and Colin Jenkins (armorer).  All three men rendered a crisp, professional salute to their pilot.

"Good to have you back, sir," said Gordon.

"Great to be back, guys.  How's the new machine?"

"She's a dream, actually," said Gordon.  "I hope you can take better care of this one than the last three.

They laughed.

"Hope you are no worse for the wear, sir," Aircraftman Toldson added.

"I'm about to fly my new airplane, Jordan.  Nothing could be better."

"Brilliant."

"Let's get her turned up," Gordon commanded.

"She's fully armed and boresighted as you like it, Mister Drummond," added Aircraftman Colin Jenkins, as Brian descended into the cockpit.

"Thanks, Colin."

The new machine smelled like it just came out of the box.  Everything from the black painted background of the instrument panel to the white labels of the various switches glistened with brightness in the mid-day sun.  The seat and harness had not yet faded from the Sun's rays.  The broad smile on his face made his facial muscles ache.  The access door closed.  Brian stepped through his cockpit checks.

The sounds, smells and feelings of the new Spitfire coming to life gave Brian a jolt of adrenaline, as he prepared for flight.  Within minutes, the three fighters were airborne.  Beamish immediately turned toward the north, the opposite of their normal path.  They checked in with Sector Control, and then spread out to go through a common set of engagement set-ups for mock aerial combat.

The flying progressed to more difficult encounters and maneuvers that strained both man and machine.  In between events of high g's when the aircraft groaned against the enormous maneuvering forces, Brian's head began to pound as his still bruised brain responded to the significantly higher circulatory system pressure his heart produced to keep him conscious during the maneuvers.  He ignored the warning signs amid the excitement of mock aerial combat with one of the world's fastest, most sophisticated fighters.

By the time they returned to Middle Wallop, Yellow Section returned with the tale of climbing to 30,000 feet in an effort to engage a high-altitude reconnaissance aircraft.  They could not quite reach the specially configured

Ju86P, but fired several bursts from long range without apparent effect.  The No.609 Squadron pilots speculated about the differences and capability of the new aircraft, or at least new to them.  The intelligence people of James Royster's section gathered as much information as they could plus retrieved the gun camera films from all three Yellow Section airplanes.

"How was your go?" 'Waggle' Davies asked Brian, as their story thinned.

"Excellent," Brian lied.  His headaches returned although not as severe as two days ago.  He felt dizzy and slightly nauseous, but knew he could not report it, if he wanted to keep flying.  "The new machine turned like a top but 'Jackstay' had the eval."

"Our young American buck flew the pants off us," Beamish lied.  Brian knew he flew well but certainly not with the intensity he was accustomed to using.  'Crazy' as well as 'Jackstay' were two of the best, most aggressive pilots in the squadron.  It was tough to beat them even on a good day.  "Fortunately, 'Hunter' kept his thumb off the trigger button, or we would be dead meat."

Squadron Leader Darling asked Flight Lieutenant Beamish to join him in his office where they undoubtedly conferred on a more personal level, regarding Brian's readiness for combat.  He knew he did reasonably well, but wondered whether Roger Beamish detected dullness of the edge in several areas.  He waited for the result.

"You guys missed the good stuff," said 'Red' Burns with the loud, strong voice.  "Who has the message?"

"Corporal Warren," shouted 'Sparky' Morrow into the Dispersal building, "if you would be so kind, please bring the invasion alert message."

"Invasion alert?" asked Kradilcek.  "Are not those damn Germans going to stop chasing me?"

They all laughed as they passed the single piece of paper among the newcomers.

---

## SECRET

```
DATE   5 AUGUST 1940
FROM   HEADQUARTERS FIGHTER COMMAND
TO ALL GROUPS STATIONS AND UNITS
RESEND
DATE   4 AUGUST 1940
FROM   WAR CABINET
TO WAR OFFICE ADMIRALTY AND AIR MINISTRY
SUBJECT INVASION ALERT
BEGIN
```

```
THE ENEMY CONTINUES TO BUILD FORCES ALONG
SOUTHERN COAST OF THE ENGLISH CHANNEL BREAK
MOTORISED BARGE TRAFFIC ALONG COAST IS AT
UNPRECEDENTED LEVELS BREAK RAPIDLY INCREASING
AERIAL BOMBARDMENT ACTIVITY CAN BE EXPECTED
FOR ALL CHANNEL PORTS AND SHIPPING AS WELL AS
MILITARY AND INDUSTRIAL TARGETS IN SOUTHERN
REGION END
BEGIN
FROM THE PRIME MINISTER THE LAND FORCES AND
HOME ARMY ARE MAINTAINED PRIMARILY FOR THE
PURPOSE OF MAKING THE ENEMY COME IN SUCH LARGE
NUMBERS AS TO AFFORD A PROPER TARGET FOR THE
SEA AND AIR FORCES BREAK EVERY EFFORT MUST BE
TAKEN TO ENGAGE ENEMY AIR AND NAVAL FORCES TO
PREVENT AN INVASION ATTEMPT END
BEGIN
FROM FIGHTER COMMAND
ALL UNITS MUST BE BROUGHT TO HIGHEST READINESS
BREAK ALL PERSONNEL SHOULD BE INFORMED
REGARDING THE IMPORTANCE OF THE NEXT FEW
WEEKS BREAK THE KINGDOM AND COMMONWEALTH ARE
DEPENDING ON US BREAK GOD WILLING WE SHALL
PREVAIL
END
```

### SECRET

---

"A proper target, ay," Beamish said. "The man has a set of bollocks, doesn't he?"

"He is practically begging them to come get it," said 'Boxer' Stockard.

"Like you, ay, 'Boxer?' Lure them into the lair, then bash the crap out of them."

"I suppose," responded Stockard.

Brian as well as several of the other pilots had watched Roland Stockard box in a couple of No.13 Group intramural boxing matches during the Phony War when the squadron was still at RAF Drem near Edinburgh. Everyone who watched him box marveled over his skills in mixing offense with defense. Roland could coax a bigger, stronger man into pressing the attack a little too quickly, and then administer a surgical blow. He enjoyed a reputation as a graceful, masterful technician with just enough sting to win.

"I wonder if the Germans are really going to invade?" asked 'Slim' Koenig with a measure of apprehension that he did not mask.

"Let me tell you," interjected Janus Kradilcek. "I have seen this too many times. In my country, then Poland, and then France, they go quiet before the attack. They are coming. We best make ready."

The group remained silent for several minutes as each of them contemplated Janus' vision and admonition. Brian's headache subsided sufficiently to allow images of German tanks rumbling across the rolling countryside of Hampshire toward them. The mire of death and destruction of Northern France and Belgium 25 years ago, as Malcolm Bainbridge had recounted, provided the tempering heat required to make the metal stronger. Brian felt the strengthening of resolute determination coursing through his body. He was not alone.

"Bring the bastards on," Morrow said, breaking the contemplative silence. "We'll give 'em a clot like our 'Boxer' does."

The group broke out in rolling laughter with the bravado. They were not particularly prone to take themselves nor the dire situation seriously. The pilots possessed a perverse form of realism, some did call it fatalism, that allowed them to laugh at danger, to make jokes about the threat. They lived on the edge, and they enjoyed it in a strange, mysterious way. Images of the beleaguered Spartans at Thermopylae enabled humorous comparisons to their situation.

The telephone rang bringing all of them back to the immediacy of the present. Brian instinctively looked south as if he might see hordes of dark, gray, German bombers heading toward them. The sky was nearly perfect. They would use the good weather to initiate their attack, but why late afternoon. Squadron Leader Darling returned to inform 'B' Flight of yet another targetless patrol. The routine mission seemed so anti-climactic after all the strong talk. They continued through two sorties waiting for a raid that did not come. Brian and several of the others remembered Kradilcek's admonition and wondered aloud what the next few days would bring their way.

———

*Tuesday, 6 August. 1940*
*No. 10 Downing Street*
*Whitehall, London, England*
*15:30 hours*

Assistant Private Secretary John Rupert 'Jock' Colville knocked twice on the Prime Minister's office door and opened it to partially enter. "Sir Henry has arrived for your scheduled private meeting, sir," he said.

Without looking up from the report before him on his desk, Prime Minister Churchill responded, "Thank you, Jock. Please show him in."

Sir Henry Thomas Tizard, KCB, FRS, was the former chairman of the Air Ministry's Committee for the Scientific Study of Air Defence and the Advisory Committee for Aeronautics, and was now unemployed after his sudden resignation after a particularly contentious meeting with the War Cabinet on the 21$^{st}$ of June.  Sir Henry had been instrumental in numerous research and development projects, focusing the scientific and engineering communities on the modern air defense dilemma.  His education at Oxford University concentrated on mathematics and chemistry.  King George VI recognized Tizard for his scientific contributions to the kingdom by granting him a knighthood in 1937 – Knight Commander, The Most Honorable Order of the Bath (KCB).

"Sir Henry Tizard, sir," announced Colville and held the door open for the accomplished scientific administrator to enter.

"Good afternoon, Sir Henry," Churchill said as he extend his hand.

"Good afternoon to you, Prime Minister," answered Tizard, as he shook hands with Churchill.

"Thank you for meeting with me." Winston motioned for Tizard to take the plump, leather chair across the small table.  "Would you care for a whiskey or tea?"

"No, thank you, sir."

Churchill nodded to Colville, who poured two fingers of his favorite Johnny Walker Black Label Scotch whiskey in a tumbler, handed the glass to the Prime Minister, and then departed and closed the door behind him.

"I do not have much time, so please forgive me for jumping directly to it." Tizard nodded his consent. "I expect we shall formally approve the technical exchange program with the Americans at War Cabinet this afternoon, in less than an hour from now.  I know we have had our disagreements, but I must assure you I have nothing but the highest regard for your keen mind and exceptionally productive administrative skills in an often factious scientific arena."

"Thank you, sir." Churchill nodded.

"On behalf of the King, I would like you to lead the team for this vital mission."

"Given past events, I am afraid I must ask if you are certain of this assignment?

"I understand, Sir Henry.  You have many proponents inside and outside His Majesty's Government, including me.  Short answer, yes, I am absolutely and unequivocally convinced you are the ideal man to lead the technical exchange team."

"What of the Professor?" asked Tizard, referring to Churchill's scientific advisor and confidante Frederick Alexander Lindemann – former professor of physics at Oxford University – Sir Henry's one-time friend and lately usual antagonist.

"To be frank and candid, the Professor energetically advanced your chairmanship for the technical exchange team. You two may disagree on specific issues, but there should be no doubt of his utmost respect for your expertise and accomplishments."

Sir Henry gazed into Churchill's eyes for more than a few seconds, as if searching for some hidden clues of sincerity and authenticity. "I am certainly available. How soon do you expect this to begin?"

"As I am sure you are quite aware, the Germans have begun preparations for their assault. They must gain air superiority in advance of their invasion attempt of the Home Islands. An invasion operation is expected anytime before the onset of autumn weather. Thus, time is of the essence. I am afraid the timing is as soon as possible."

"Has the list been finalized and agreed?"

"Yes. It will be formally approved this afternoon."

"Has the team been identified and alerted?"

"Yes. We only need the leader to bring it all together. I know Sir John has contracted for shipment by the end of this month. Lord Lothian has made arrangements for lodging, storage and office space. You will need to meet with Sir John to get details and with your team before the week is out. If you agree to this critical assignment, we shall leave you to refine the details and execute the exchange."

Sir John was Home Secretary and Minister of Home Security the Right Honorable Sir John Anderson, GCB, GCSI, GCIE, PC, Member of Parliament for the Combined Scottish Universities, who was not affiliated with a political party, and in this instance served as the ministerial sponsor for the technical exchange program.

Lord Lothian was actually Philip Henry Kerr, Kt, CH, PC, DL, 11th Marquess of Lothian, and the well-regarded British Ambassador to the United States of America.

"Since I have been out of discussions for the last six weeks, are we to expect any *quid pro quo* with the Americans?"

"No. This exchange will be unilateral – an insurance policy, if you will. We want the Americans to have as much of a head start as we can provide . . . should the German invasion be successful," Churchill said with solemnity. "There must be no expectation of reciprocity."

"Very well."

"That said, if the information provided by Lord Lothian is correct, which I have no doubt he is, then the Americans appear to be eager to share their defense research and development work to optimize our technical capabilities – an allied collaboration of the most intimate kind."

"Then, so it shall be."

"You agree?"

"Yes, I would be honored to lead the exchange team and do my part to ensure the defense of the kingdom and the empire."

"Excellent."  Churchill stood, extended his hand to Tizard.  "Congratulations, Sir Henry, and thank you for serving the King.  Now, if you will forgive me, I have a very busy schedule this afternoon."  After shaking hands, the Prime Minister motioned toward the door.  "Godspeed and following winds, Sir Henry."

"Thank you, sir."  Tizard closed the door behind him.

Prime Minister Churchill returned to his desk, checked several papers, and then pressed one of the levers on his desk interphone box.

"Jock, would you ask Sir Edward to join me as soon as he is able?"

"Yes, sir."

*15:50 hours*

The Prime Minister needed to coordinate the last minute details for the War Cabinet's evening meeting.  The message he received from the President of the United States buoyed his will, confidence and optimism.  Although it did not commit the weight and capacity of America to their current conflict, it was certainly the most encouraging message to date.  Winston knew the War Cabinet needed encouragement despite their outward strength and support.  He just wished he could share the communication with the nation.  Morale remained unusually high despite all the invasion preparations and warnings, but everyone could use a shot in the arm like this one.

Besides the encouraging news from across the Atlantic, the situation in and around the Channel area did not offer much evidence for optimism, and yet, Winston knew he had to find the hope in any situation.  For the strangest reason, while he waited for the Cabinet Secretary to arrive, his thoughts drifted off to a time forty years earlier when he found his own depression and remorse, the Black Dog as he called it, during his harrowing escape from a prisoner of war camp during the Boer War in South Africa.  All alone in hostile territory, he had to find hope to fight his own demons and trust a few benevolent strangers during his escape.  Winston knew from his many and varied experiences that you always had a choice . . . to focus your attention on how bad things were, or latch onto those few, small kernels of brightness among the morass of dismal conditions.

The knock at the door meant Sir Edward Ettingdene Bridges, KCB, MC, FRS, Secretary to the War Cabinet, had arrived.  The usual amenities between colleagues centered on the unusually dreary and windy weather on this mid-summer day.  The storm seemed to be rather large and slow moving offering the possibility of protracting the respite from aerial attacks.  While they said the words, neither of them believed them.

"Nonetheless, Ed," Winston said, wanting to change the subject, "I wanted to finalize the agenda for this evening's meeting."

"Certainly, Prime Minister."

"We appear to have some good news as well as some not so good news.  I should think it better to end on a good note.  In that vain, I would like to share with the Cabinet this note I received this morning from President Roosevelt."  Winston handed Sir Edward the single piece of paper.

---

### TOP SECRET

57

August 4, 1940

```
FROM: POTUS
TO: Former Naval Person
     We share your concerns indicated in
your most recent message.  Our Navy continues
to watch the situation in the Atlantic, with
considerable worry, I might add.  I have asked
Admiral King to look into the possibilities,
which might be available to us regarding your
request for destroyers.  We shall endeavor to
respond as quickly as we possibly can.  We are
quite sensitive to the urgency of the present
situation.
     We shall also continue to ship rifles
and other munitions to you, God willing.  We
can only pray our meager assistance will be
sufficient to sustain the British people during
this time of crisis.
     Recently, I decided to reread
Shakespeare's Henry V as the words of King
Harry came to me several times of late.
"This story shall the good man teach his son,
```

```
and Crispin Crispian shall ne'er go by
from this day to the ending of the world
but we in it shall be remembered,we few, we
happy few, we band of brothers.
For he today that sheds his blood with me
shall be my brother; be he ne'er so vile,
this day shall gentle his condition.
And gentlemen in England now abed shall think
themselves accursed they were not here,
and hold their manhoods cheap whiles any speaks
that fought with us upon Saint Crispin's Day."
        While very few of us have the honor
to shed our blood with you, you should know
that we are with you in spirit, and we shall
not rest in our efforts to provide support
and encouragement.  We shall pray for your
deliverance from this evil that surrounds you.
Godspeed and following winds.
        FDR
```

## TOP SECRET

---

"A most encouraging message, sir."

"I thought so, as well.  However, words can do little to hold back the tide."

"You sell yourself short, Prime Minister."

"What do you mean?"

"You may not be fully aware of public opinion.  Since you joined the government last fall and began your radio addresses, the people as well as the military have derived substantial strength from your words alone.  You must realize how important your messages are to the people."

"I am a simple politician and orator trying to perform the duties thrust upon me."

"Nonsense, Winston.  I have known you long enough.  You are the roar of the lion, and they shall know his power."

Winston smiled to himself.  Sir Edward, although he chose not to enter professional politics, was the consummate politician. Winston took great pride in his oratory skills, but did not take pleasure in flaunting them.  He decided to ignore the compliment.  "I trust you will agree we should conclude this evening's meeting with the President's message."

"Excellent."

"What else do we have on the agenda?"

"As is usually the case, we should begin with the intelligence briefing. The indications are becoming more pronounced. The home defense preparations should follow, and then the situation in the Atlantic. You wanted the technical exchange program to be decided at this meeting. And, as you suggested, we can end with President Roosevelt's encouraging message."

"Agreed."

"Anything else, sir?"

"I should like to convene the Defense Committee an hour prior to the Cabinet."

"As you wish, sir. As you may know, Mister Chamberlain has taken ill, and, I am afraid, is bedridden for the next several days."

The Right Honorable Arthur Neville Chamberlain, FRS, Member of Parliament for Birmingham, Edgbaston, continued to serve on the War Cabinet despite suffering from terminal intestinal cancer that his attending doctors chose to keep hidden even from their patient. Without knowledge of Chamberlain's illness, Churchill chose to retain his predecessor as Lord President of the Council. The former prime minister remained the leader of the Conservative Party in the House of Commons.

"Yes. I have sent my sympathies."

"Mister Alexander had a family emergency and shan't be with us this evening," Sir Edward said, referring to First Lord of the Admiralty the Right Honorable Albert Victor 'A.V.' Alexander, PC, Member of Parliament for Sheffield, Hillsborough, a member of the Labour Party, successor to Churchill as First Lord, and ministerial leader of the Royal Navy..

"Nothing serious I hope."

"No, sir. His daughter seems to have taken a header off her horse."

"Oh, dear. I shall telephone his family after the Cabinet meeting."

"All the others should be available."

"Excellent. Please ask Admiral Pike and Colonel Menzies to join us for the Committee meeting."

Secret Intelligence Service (SIS or MI6), Director General Colonel Stewart Graham 'C' Menzies, DSO, MC, carried the traditional 'C' moniker as chief of MI6 in honor of the SIS founder Captain Sir George Cumming. Menzies ascended to the head of MI6 the previous year with the passing of Admiral Sir Hugh Sinclair. Director of Naval Intelligence Vice Admiral Sir Geoffrey Ian 'Jumper' Pike, KCB, DSC, had held his post since 1935. His shipmates aboard the destroyer HMS *Spitfire* during the epic Battle of Jutland in the Great War had affectionately given him the nickname 'Jumper' for his heroism and sacrifice, for which the King awarded him the Distinguish Service Cross for his conspicuous gallantry in combat action.  Both men were well known and trusted by Churchill.

"As you wish.  Anything else, sir."

"No, that should do, Sir Edward."

The Cabinet Secretary departed to perform his duties.  Principal Private Secretary John Miller Martin took the opportunity to return several papers that had been typed from notes earlier in the day.  The exchange of information blurred the passage of time.  The knock at the door changed his mood.  John Martin opened the door slightly and positioned himself partially in the space.

"Mister Attlee to see you, sir," Martin announced.

Winston nodded his round, balding head and stood to greet Clement Richard Attlee, Member of Parliament for Limehouse – Lord Privy Seal and Leader of the Labour Party.  He glanced quickly to the wall clock.  "Good evening, Clement."

"Good evening, Winston.  How are you holding up this stormy day?"

"Rather nicely, I should think, and you?"

"Better than Neville.  I suppose you have heard?"

"Yes.  He seems to be in a descending spiral to coin a term from the aviator's lexicon."

"Tragic actually, but inevitable nonetheless."  Attlee paused to allow silence to define the demarcation in topics.  "I understand from Sir Edward you have called for a precursor Defense Committee meeting."

"Yes."

"Something new?"

"Actually, the news builds up by the minute.  The situation across the Channel is looking much worse by the day.  I would like to discuss our options within the Defense Committee and present a more cogent set of alternatives to the War Cabinet this evening."

"What does it look like to you?"

Winston understood the context of the question as well as the proper answer.  "If the information we have from aerial photography as well as other sources is accurate and reasonably representative, the bastards should be ready on their part within a few more weeks."  Winston knew what the Germans considered important.  "The key lies in air cover.  If Fighter Command can maintain sufficient superiority over the southern approaches, we can hold them at bay.  If they crack the fighter defenses, then we shall be faced with fighting our pitched battle in the villages and meadows of Kent, Sussex and Surrey."

"Do you think Fighter Command can hold up to the task?  After all, there are now much less than a thousand, pink-cheeked, young men out there.  If I recall the order of battle presented by Archie and Air Chief Marshal Newall, they have two to three times the number of fighters and maybe five times the number of bombers as the RAF."

"Your memory is close enough."  For him, there was no, if.  There was no choice.  Failure simply would not be permitted.  While the Germans might very well be able to accomplish an invasion of the British Isles, they would not be able to sustain their occupation of even a portion of the island.  Winston debated with himself whether he should articulate his innermost thoughts.  The public image and the confidence of his colleagues had to be maintained.  If any of them lost hope and the will to succeed, they were all doomed.  "I have pondered the same question for many months now.  One friend and colleague to another, the future of our great nation and empire rides on the shoulders of those dauntless fighter pilots.  While their spirit is unquestionably up to the task, the weight of the Nawzee war machine may prove undeniable.  However, my public opinion is, those young men shall rise to the occasion as Sir Francis Drake, Admiral Lord Nelson and John Churchill, the 1st Duke of Marlborough, did in their time.  We shall prevail."

Churchill loved to nearly slur the pronunciation of the Nazi label for members of the *National Sozialistische Deutsche Arbeiter Partei* (National Socialist German Workers Party) – the foundation for and means by which *Reichskanzler* Adolf Hitler dominated German society and its armed forces, the *Wehrmacht*.  All too often, the Nazi label was erroneously applied to all Germans, since only a minority of the populace were actually members of the political party.

"You never cease to amaze me, Winston.  I have never known a man of your demonstrated courage and fortitude.  As you say then, so it shall be.  We shall prevail."

"We must stand together, Clement."

"I know.  I shall do my part to ensure we have solidarity through this trial."

"Excellent."

"Clement, from my perspective, everything points toward the next month or two, at least until the onset of foul weather this autumn, as the watershed for one side or the other.  If Hitler and his Nawzee henchmen do not succeed this summer, we shall gather strength and gain the full support of the Americans.  Once that has occurred, we shall grind them down.  The next few months are critical."

"Agreed."

"Then, let's be off," Winston said, and then remembered Roosevelt's message. "Oh, my dear fellow. I nearly forgot. I received this message from President Roosevelt this morning." He handed the paper to the Lord Privy Seal.

Attlee read the message carefully several times. Winston thought he saw a glistening in his colleague's eyes. Attlee cleared his throat. "Magnificent. I had forgotten that passage. There could be no more appropriate set of words." He cleared his throat again, obviously moved by the note. "You intend to share this with the Committee and Cabinet?"

"Yes."

"Brilliant, Winston. I do believe as you, my friend."

The two political leaders remained in opposition. Yet, they set their differences aside in the coalition government. Attlee had become one of Churchill's staunchest supporters.

*17:30 hours*

Sir Edward Bridges knocked and entered to inform the two political leaders that the Defense Committee, less Chamberlain and Alexander, was present. After their acknowledgment, the Cabinet Secretary closed the door behind him.

"Would you mind if I observed your Defense Committee meeting?" asked Attlee.

"By all means, you are most welcome. Shall we?" Churchill said, motioning toward the door.

In addition to the Prime Minister and Lord Privy Seal, the other attendees were:

-- for the Royal Navy, First Sea Lord Admiral of the Fleet Sir Alfred Dudley Pickman Rogers Pound, GCB, GCVO, GBE;

-- for the British Army, the Right Honorable and Gallant Robert Anthony Eden, MC, PC, Member of Parliament for Warwick and Leamington – Secretary of State for War, Deputy Leader of the Conservative Party, and ministerial leader of the British Army; and Chief of the Imperial General Staff General Sir John Greer Dill, KCB, CMG, DSO; and,

-- for the Royal Air Force, the Right Honorable Sir Archibald Henry Macdonald 'Archie' Sinclair, Bart, PC, CMG, Member of Parliament for Caithness and Sutherland, 4th Baronet of Ulbster, Secretary of State for Air, Leader of the Liberal Party, and ministerial leader of the Royal Air Force, and Chief of the Air Staff Air Chief Marshal Sir Cyril Louis Norton Newall, GCB, CMG, CBE, AM; as well as the four Cabinet Secretariat service members:

– Major General Sir Hastings Lionel 'Pug' Ismay, KCB, DSO;

-- Captain Angus Dacres Nicholl, RN;

-- Lieutenant Colonel Vivian Dykes, Royal Engineers; and

-- Wing Commander William Elliott, RAF.

As the Prime Minister had requested, two of the intelligence chiefs were also waiting – Colonel Menzies and Admiral Pike. Winston allowed a brief thought and prayer for their ill colleague before he opened the meeting. It was the beginning of a long night as they struggled with the defense of the United Kingdom and her Empire.

"Colonel Menzies, if you would be so kind to provide us the latest information on the situation across the Channel?"

"My pleasure, Prime Minister. As of mid-day, the build-up of transport vessels has been progressing rapidly. The German field engineers have been feverishly clearing the battle damage in all of the Channel ports from Ostende to Le Havre, both for collection of their transport and preparations for loading those vessels. Based on their progress since the French surrender . . ."

"The French might quibble with your use of the word surrender," interjected Churchill.

"Yes, well, shall we say, once the French stopped fighting . . . we estimate they will be ready to make an attempt within two weeks. They have garrisoned five infantry divisions and two armor divisions along the coast and have clearly been conducting training exercises. We have provided targeting information to the Air Ministry."

Churchill looked to Sir Archie, expecting a response.

Sinclair took the cue. "We have had limited success with night bombing. We have not undertaken daylight bombing raids, as we husband our fighter resources."

"Are we to give them a free pass?" asked Winston, with a heavily sarcastic tone.

"No, sir. However, we are faced with precisely the same dilemma we did in May. We have precious few fighters, and even more precious few pilots to fly those aircraft we have. The risk to the home defense is simply too great fighting over there. I do not know how many pilots we have successfully recovered and saved from the Channel, but it is more than a few. We would be in a far more precarious position if those pilots had been captured in France, or even worse shot by the Germans. Bomber Command will do its best to bomb the Channel ports and staging areas at night, without a fighter escort. That is simply how it is and how is must be."

"Hmmm," grunted Churchill.

"Since I seem to have the floor at the moment, Fighter Command has been holding their own over the Channel shipping."

"We lost another destroyer just yesterday," Sir Dudley said, taking his turn to interject.

"Yes, well, we do not have sufficient fighters to maintain around the clock air cover over every ship."

"We all understand that, Sir Archibald," responded Admiral Pound. "I simply noted the losses to Channel shipping may well force us to close the Channel to merchant shipping."

"The enemy has become quite adept at mixing high and low altitude attacks. The Chain Home Low system gives us perhaps ten minutes warning to our southern beach line by low altitude intruders – less than one hundred feet – but that warning time decreases rapidly to the nautical horizon, and offers no warning to ships that transit farther out in the Channel. The Admiralty is managing surface traffic to keep shipping within Chain Home Low coverage to the greatest extent possible."

"We understand the physics, Archie," the Prime Minister said. "I think we all share the same frustration that we are unable to do everything at once to protect what is ours."

Sinclair nodded his acknowledgment. "I think we have sufficient evidence that the German air attacks on Channel shipping is the prelude to their main assault." Archie looked to Menzies, as a few other attendees did as well. 'C' nodded his head in agreement. "The main assault will come shortly and will be directed specifically at Fighter Command in the southeast. That phase, when it comes, will be the final contest for air superiority. It is that phase we must win to baulk their Channel crossing attempt, which you have noted many times must come before summer is out."

"I expect to brief the War Cabinet this evening on our situation and our defense plan. We are rapidly approaching a critical juncture. Time will become even more precious. We have focused on the Air Force for quite some time now. I would like to hear from the Admiralty, and then the War Office."

The First Sea Lord leaned forward. "The First Lord asked me to convey his apologies for not being able to attend. His daughter had a serious accident and was taken to hospital."

"Thank you, Sir Dudley. Please pass our best wishes for Beatrix's speedy recovery."

"Yes, sir. I would be honored to do so. To the naval situation . . . Admiral Ramsay has continued to manage the Channel shipping traffic as well as the rotation of our destroyer squadrons on escort duty."

"Now, Sir Bertram," Churchill interjected, noting the recent knighthood bestowed by the King on 7.June.1940, upon Vice Admiral Sir Bertram Home

Ramsay, KCB, MVO – Knight Commander, The Most Honorable Order of the Bath – for his extraordinary leadership during the Miracle of Dunkirk and the saving of a third of million vital soldiers from the surrounded beaches of France.

"Yes, indeed, and well deserved, I must say. Sir Bertram has advised the cost may exceed the benefit, and it may well be time to abandon pressing the Channel with merchant shipping."

"Is he now advocating we give up?"

"No, sir. We have two squadrons in and around the Channel at any given time. We also have two additional squadrons being rearmed, refitted and rested for duty in the Channel and approaches. We are committing a large portion of our total destroyer force to this effort, rather than trans-Atlantic convoy escort or submarine hunting. We have other, safer, available ports in the Kingdom. Furthermore, if the intelligence estimates are correct, the Navy will soon need all the maneuvering room it can gain to deal with an invasion force without interference from merchant shipping. We also have two full, heavy cruiser squadrons fully armed and on dispatch alert should we see signs of embarkation by the Germans. Perhaps Admiral Pike would care to update the Defense Committee."

Sir Geoffrey cleared his throat. "Our field agents have accounted for all of the German capital ships, and they are maintaining watch at considerable risk to their safety. There are no signs they are making ready for sea, although they likely could do so within a few days, if ordered to put to sea for battle. The Germans have sortied a half dozen U-boats from Wilhelmshaven and Hamburg; in the last week; however, we suspect they are more likely redeployments to Norway and the occupied Atlantic coast ports of France, probably Brest or Saint-Nazaire. We have seen no evidence or even hint those boats or any others would risk the shallows and confines of the Channel."

"Thank you, 'Jumper.' I know your lads will keep an eye on things, but keep them safe. We need their eyes and insight," the Prime Minister offered.

"Yes, sir."

A knock at the door produced silence and turned every head toward the door. Sir Edward entered, went directly to the Prime Minister and whispered in his ear, "Sir, the War Cabinet is gathering for the scheduled meeting in 15 minutes. What should I tell them?"

Churchill mumbled, "We shall be done shortly."

"Very well, sir." Bridges departed and closed the door behind him.

"Let us hear from the War Office."

"General Brooke has done and continues to do a masterful job preparing our beach and land defenses," Anthony Eden reported, referring to

Lieutenant General Sir Alan Francis Brooke, KCB, DSO, who was also knighted by the King on 14.June.1940, in recognition for his exceptional defense of the Dunkirk perimeter during Operation DYNAMO. Sir Alan now served as General Officer Commanding Home Forces and was responsible for the land defense of Great Britain. "Our efforts to refit, reconstitute and redeploy units to defensive positions are slowing, as we have distributed all the weapons sent by the Americans. We need more."

"I can assure you, President Roosevelt is working hard to that end," answered Churchill with some degree of hope mixed in with facts. "We have not yet received notice of the next shipment."

"We will put them to good use as soon as they arrive."

"I have visited several of the defensive lines as well as the potential landing beaches," said the Prime Minister. "I know they have done well with the time and resources we could give them. Is there anything else we can or should be doing?" They all either remained silent or shook their heads in the negative. "Very well, then, we have the War Cabinet stacking up on us. Anthony, Archie, if you both would sit in, it would be most helpful."

"Should I remain for Mister Alexander?" asked Admiral Pound.

"Not necessary, Sir Dudley."

"Very well, sir."

The Secretariat officers remained, as was the custom. The intelligence and combined chiefs departed, and the remaining members of the War Cabinet entered and took their seats – Arthur Greenwood, Member of Parliament for Wakefield, Minister without Portfolio and Deputy Leader of the Labor Party, and Edward Frederick Lindley Wood, KG, GCSI, GCMG, GCIE, TD, PC; 3rd Viscount Halifax of Monk Bretton, Secretary of State for Foreign Affairs; Conservative Party and Peer in House of Lords, and popularly known as Lord Halifax. Cabinet Secretary Sir Edward Bridges also joined them in the Cabinet Room. Since the technical exchange program was on the agenda, Home Secretary Sir John Anderson also joined them.

*19:00 hours*

Prime Minister Churchill did not wait for the usual formalities or for the door to be closed. Winston recounted the Defense Committee discussions to summarize the nation's current defense state. Attlee heard it directly, and Greenwood and Halifax were sufficiently informed to appreciate events.

"Truth or consequences time will soon be upon us. I believe we can all agree, it is time to issue an invasion alert." The three other cabinet members nodded their heads in agreement. "Sir Edward, if you would be so kind, please prepare a simple invasion alert message to His Majesty's Government, all ministries, to the effect that we should expect heavy aerial combat and bombardment of land facilities as the enemy prepares for an invasion attempt as early as two weeks from now – every man to his post. The time to stand for King and kingdom is here. The next two months will most likely decide this war. Either we stop the enemy now, or we shall join our neighbors for the descent into a long dark night. Let us not prove ourselves wanting."

The room remained silent. Only the scratching of Sir Edward's fountain pen on his paper tablet could be heard. The solemn moment lasted far longer than was characteristic for Winston Churchill. Bridges completed his notes and waited. Churchill raised his head, as if from a silent prayer.

"One last item that we must get to tonight – the technical exchange program. The list of items was finalized and agreed last week. The time for my indecision has passed. After the recommendations of many, I met with Sir Henry Tizard earlier this afternoon. He has agreed to put bygones aside and take the helm as the leader of the exchange team, whose membership has been agreed as well. We have a quorum of the War Cabinet and I would like final approval and go ahead from this very important insurance policy. As this is another historic moment, Sir Edward, if you would be so kind, please poll each member of the War Cabinet and record their vote."

"Very well. Mister Attlee?"

"Ay."

"Mister Greenwood?"

"Yes."

"Lord Halifax?"

"Yes."

"Mister Churchill?"

"Yes."

"I record four affirmative with two absent. The Technical Mission to the United States of America is approved."

"Thank you, Sir Edward, and thank you gentlemen. Now, in the interest of time, I gave Sir Henry provisional approval to organize a meeting of the team as soon as possible with the assumption he was to be the team leader, unless he heard to the contrary tonight. So, Sir Edward, please get word to Sir Henry tonight that the War Cabinet has approved the mission and his leadership of the team. I expect him to finalize arrangements with the team in the next few days and notify us of their final execution program. Lord Halifax, I think it quite appropriate to notify Lord Lothian of the decision we have taken this evening and that details will be forthcoming from Sir Henry's team."

"It will be my pleasure," answered Halifax.

"Sir John, my congratulations, perhaps the congratulations of the entire War Cabinet, for the yeoman's job bringing this whole program together."

"Herding cats, actually." Anderson said.

Everyone laughed. It was an extended, cathartic laugh that released the tension built up in all of them over the last few days. They had laughed so rarely. It felt good to all of them.

"Quite so. Please convey our appreciation to all those involved for a job well done." Anderson nodded his agreement. "Now, we must hope and pray this insurance policy is never called, and that our American cousins receive our magnanimity in the spirit it is given. Sir John, if you would, please continue as the ministerial shepherd for this program, and ensure the team receives whatever resources and support they need for the prompt and successful execution of the plan."

"As you command, Prime Minister."

"There it is. The deed is done. Is there any additional business we need to conduct this evening?" Again, silence and negative gestures. "Excellent. Then, Sir Edward, please mark us concluded. Good evening, gentlemen." They departed in an upbeat mood. Churchill checked with John Martin and Jock Colville to ascertain there were no urgent actions, and then proceeded upstairs to the apartment and what he hoped to be a quiet supper with his wife, Clementine Ogilvy Spencer-Churchill née Hozier.

———

*Wednesday, 7 August. 1940*
*Headquarters, Fighter Command*
*Bentley Priory*
*Stanmore, Middlesex, England*

Group Captain John Henry Randolph Spencer, DFC, was already deep into the day's paperwork, that never ending stream from reports, messages, letters, notices and other myriad forms of document that was the business of the staff secretary to Fighter Command and principal administrative assistant to Air Officer Commanding-in Chief, Fighter Command, Air Chief Marshal Sir Hugh Caswall Tremenheere Dowding, GCVO, KCB, CMG.

Dowding had served as Air Officer Commanding-in-Chief, Fighter Command, since its formation in July 1936, and had overseen the development and operational integration of the new monoplane fighters – the Supermarine Spitfire and the Hawker Hurricane. He had also advocated for and shepherded the deployment of the Radio Direction Finding and ranging system known as Chain Home, along with the command and control system to utilized the new detection equipment. If one man could be credited with the preparedness of the Air Defense system of Great Britain, that man would be Sir Hugh Dowd-ing – the father of Fighter Command. Dowding was affectionately known as 'Stuffy' by his men – his chicks as he called them.

The layers of low clouds over the English Channel and most of South-ern England kept the German air activity confined to long range, high altitude reconnaissance flights over the midlands and Scotland. Although Operations was not the beehive of movement experienced over the last month, the staff efforts, from Group Captain Spencer's perspective, held an uncommon intensity all day. The brief respite from air defense operations had given him the break he needed, and he took it. John found himself quietly thanking the weather and the Germans for the few moments.

John had managed to take the previous evening off and take his wife – Mary Elizabeth Ann Spencer née Armstrong – out to supper at their favor-ite restaurant. They even found some time and energy for the more intimate pursuits within the marriage. It had been too long for John, and he knew the strain of the war and his commitments to the Nation and Empire continued to corrode the bonds of their relationship. At least this time, Mary made overt attempts to bridge the growing gap between them. He wanted to spend more time with her, but he always rationalized that she could wait . . . the Nation's defense could not. He could only pray he made the correct judgment.

The telephone rang. John wished his assistant had not been ill this day. "Spencer," he said.

"John, this is James," responded Air Commodore James Hogan, DSO, Chief of Air Intelligence, Fighter Command.

"Good afternoon to you, commodore. What can I do for you?"

"And good afternoon to you as well. I just received several items, which might be of interest to you. Do you think you might spare a short time to join me in my lair?"

John glanced at the stack of unread papers in his basket. He knew he could not leave the headquarters building until he had dispensed with the stack. "Certainly," he answered with as much enthusiasm as he could muster.

He reluctantly rose from his desk to begin the journey down to the far corner of the underground bunker. As he exited his office, he turned in-stinctively to his assistant's desk as he always did to tell her of his destination. He was immediately reminded of her absence and felt a renewed surge of frustration. The pair of eyes in the small slot in the locked steel door identified him before the locks were commanded to give way.

"What gems do you have for us today?" John said, as he entered the office of the Chief of Air Intelligence Branch.

Air Commodore Hogan retrieved a single paper from one of several stacks. "I just received this," he said as he handed the message to John, "and thought you would be interested in it."

---

## MOST SECRET

```
DATE 6 AUGUST 1940
FROM AIR MINISTRY
TO HEADQUARTERS FIGHTER COMMAND
SUBJECT INTELLIGENCE SUMMARY NO 40 218
BEGIN
VARIOUS RELIABLE SOURCES INDICATE ENEMY
INVASION PREPARATIONS ARE PROGRESSING RAPIDLY
BREAK DETERMINED OFFENSIVE OPERATIONS AGAINST
FIGHTER COMMAND INSTALLATIONS AND ASSETS AS
WELL AS AVIATION INDUSTRY TARGETS CAN BE
EXPECTED AT ANY TIME END
BEGIN
ENEMY FIGHTER AND BOMBER RESOURCES IN CHANNEL
REGION CONTINUE TO GROW BREAK RECENT OPERATIONS
SUGGEST ENEMY WILL UTILIZE LARGE BOMBER
FORMATIONS ESCORTED BY EQUAL OR LARGE FIGHTER
FORCE END
BEGIN
ENEMY AIR FORCES WILL MOST PROBABLY FOCUS
OFFENSIVE AGAINST RAF FIGHTER RESOURCES IN A
CONCERTED EFFORT TO PRODUCE AT LEAST LOCAL
AIR SUPERIORITY OVER POSSIBLE LANDING BEACHES
BREAK SOUTHEAST CAN EXPECT TO BEAR THE BRUNT OF
THE ATTACKS BREAK EVERY EFFORT WILL PROBABLY
BE MADE TO DRAW FIGHTER COMMAND INTO DECISIVE
ENGAGEMENTS INTENDED TO DEPLETE AIR DEFENCES AS
RAPIDLY AS POSSIBLE BREAK VARIETY OF SOURCES
SUGGEST A KEY DATE FOR INVASION DECISION MAY BE
ON OR ABOUT 15TH AUGUST BREAK HIGHEST READINESS
SHOULD BE MAINTAINED
END
```

## MOST SECRET

---

"Isn't that a pleasant synopsis," said John sarcastically.

"Quite, and there's more."

"Bloody hell."

James Hogan laughed. The Staff Secretary's frustration was too obvious. "Yes, well, war is hell, isn't it, John. Operations received notice from the Admiralty we shall have a plump target in the Channel tomorrow. Convoy CW9, code named PEEWIT sails from Thames Estuary at high tide this evening for a westerly transit."

"Why can't they simply go north and avoid being a damn target."

"Now, now, John. You should know better. The Admiralty is not about to let a few bombs dictate their operations. Plus, we could always look at the bright side. While the shipping is available for bombardment, Gerry has left the aerodromes untouched. We should be thankful."

"I suppose you are correct," John said begrudgingly. "Have the groups been notified?"

"Underway, now, I should think. Bomber Command has also been directed to make an extra effort to hit as many air force targets as possible to slow them down a little. They are to concentrate on Channel area aerodromes."

"We shall see how well they do."

"Indeed."

"Anything else, commodore?"

Air Commodore Hogan retreated into his thoughts considering the mass of intelligence and related material he had seen just this day. His eyes brightened as he returned and raised his right index finger. "Please close the door," Hogan said. He waited until John Spencer sat down in the chair across from his desk again. He turned the dial on his desk safe, opened the door with the characteristic metallic clanking associated with heavy locking systems and pulled out another message. This paper had broad red stripes top and bottom that told him it was marked 'Most Secret' and required special handling. "As always, I must inform you this information is classified Most Secret, and you will understand why."

Group Captain John Spencer took the message but looked at Hogan's curious expression like a young boy holding a special secret from his younger brother. It was important.

---

## MOST SECRET

```
DATE 6 AUGUST 1940
FROM AIR MINISTRY
TO HEADQUARTERS FIGHTER COMMAND
```

```
SUBJECT INTELLIGENCE SUMMARY NO 40 219
BEGIN
MULTIPLE SOURCES INDICATE ENEMY MAY SHIFT
TARGETING FOCUS FROM CHANNEL SHIPPING TO
FIGHTER COMMAND FACILITIES AND OPERATIONS AS
EARLY AS SATURDAY 10 AUGUST BREAK EXPECT HIGH
INTENSITY AIR SUPERIORITY OPERATIONS IN DAWN TO
DUSK HOURS BREAK NIGHT BOMBER RAIDS TO ACHIEVE
MAXIMUM PRESSURE ON RAF FIGHTER COMMAND MAY
SUPPORT FIGHTER OPERATIONS BREAK ATTACKS ON AND
SUPPRESSION OF RAF FIGHTER COMMAND CONSIDERED
AN ESSENTIAL PREREQUISITE FOR INVASION PER
ALERT BY INTSUM NO 40 218
END
```

## **MOST SECRET**

"Interesting, is it not?" Hogan said finally.

"Smashing, absolutely bloody smashing. Has 'Stuffy' seen this?"

"An hour ago, as soon as I finished reading it."

"He must be pleased."

"Hard to say, actually. You know 'Stuffy.' Never shows you a face."

John Spencer considered the words once more before he handed the important message back to its guardian. "So, it shall be. This shall be the purest of contests. The lads in the cockpit are the small band of skilled and devoted knights of old."

"My, my, my, I didn't know you had this romantic streak, John."

"Well, isn't it though? While many will contribute and assist, it will all come down to the battleground in the sky and those few warriors who must face the enemy. We have all known it would come to this, but here we are. The actual battle will soon be joined."

Air Commodore Hogan reread the message himself one more time before returning it to the safety of his desk safe. "And, so it shall be," he said, repeating the words of the Staff Secretary.

"Indeed. I trust the group commanders and especially Air Vice Marshal Park have been notified."

"Shortly after Sir Hugh read the message."

"Excellent. What does the weather look like for tomorrow?"

"The trailing edge of this system will leave broken clouds at various altitudes and generally fair."

"Then, we are at the eve of the great battle."

"It has been rather great already. We seem to be at a major demarcation in their battle plans, and we shall see a new phase of the battle."

"You are quite correct, as usual, James. What other treats do you have?"

"That is it, I'm afraid. The remainder is so boring it would put you to sleep immediately."

"I should be off to my paperwork. I shall jump in with renewed vigor. I imagine we will all spend the day in the Operations Room gallery tomorrow."

"It would seem so."

Group Captain Spencer felt the change in the air like the great cat senses the change in its hunting fortunes. He also felt an excitement the aircrews could not. He wanted to tell them what was before them. He wanted them to know, to recognize the significance of what was about to happen. He also knew his words would serve no purpose. The fighter pilots would face the battle regardless of the intelligence or situation. John felt the urge to be in the cockpit more than ever before. They were now immersed in a historic event, and they could not appreciate the emotions of an old warrior watching the battle from the periphery. As the night ground on, his thoughts eventually returned to Mary. He wanted to tell her, but she would not understand. Near midnight, sleep absorbed him as he sat at his desk.

—

# Chapter 2

The first blow is half the battle.

-- Oliver Goldsmith

*Thursday, 8.August.1940*
*RAF Middle Wallop*
*Middle Wallop, Hampshire, England*
*05:45 hours*

### Week 5

The rich aroma of freshly cooked bacon mixed well with the earthy smell of coffee brightened an otherwise dark start to the day. The broken clouds and moderate breeze probably signaled a busy day for the fighter squadrons. For Brian, it was not the prospect of impending combat that darkened his day. He could not expunge the images of his nightmare. It was the most vivid dream he had had in many years.

"You look like hell," Jonathan Kensington said to his friend. "I didn't think you went pub crawling last night."

"I didn't," Brian answered quietly, as they finished their breakfast.

Brian did want to talk about his dream with his best friend, but he did not want to talk about it where other pilots might hear and open him up for the usual ridicule that accompanied such revelations. Despite several attempts by Jonathan and others to get Brian to talk, he stayed within his thoughts. He could not resist the temptation to interpret and attach significance to such a graphic dream. Was this a premonition of what lay ahead? The question kept coming to him. Was this an artifact of his concussion last week? Was this also what he had to look forward to for the rest of his life? The nightmare actually scared him.

Jonathan and Brian walked slowly across the A343 carriageway and through the main entrance gate to the airfield. Brian glanced over his shoulder several times to ensure other pilots were not within earshot. "I had a dream last night," Brian finally said softly to Jonathan, "and it scared the hell out of me."

"It was just a dream, Brian. Dreams cannot hurt you."

"This one was different."

"How so?"

"I was in London walking down the street from an Underground station late at night presumably toward Anne's townhouse."

"Brian, you are dreaming about her because of what is going to happen to her," interjected Jonathan.

"That's not it. I could hear air raid sirens as clearly as if I was standing there. Bombs began falling in the distance like a strange, weird thunderstorm.

The explosions began to move closer. I couldn't make it to Anne's place. I found a call box. I remember the shape and color, the red, deep red wood flashing in the lightning brilliant light of the explosions. I decided to call her, to tell her she needed to find shelter from the bombing as if she wouldn't have known. I was talking to her. She was frightened and asking me to help her. Bombs fell all around me while I was talking to her. Fires started in all the buildings . . . big fires, but I had to talk to her, to calm her. Then, the buildings began to collapse all around me. Before I could do anything, there was fire surrounding that call box. I was trapped. The door wouldn't open. I could feel the heat from the fire. I was trapped, and my clothes began to smoke, and then I woke up. I couldn't sleep the rest of the night."

They walked on for several yards before Jonathan said anything. "Jesus, Brian. No wonder you look so bad."

"Yeah. I keep asking myself if this is a sign."

"Dear God above, Brian, you must put those thoughts out of your mind. Dreams are just dreams. They are not premonitions. Besides, the woman is going to be executed for treason day after tomorrow, and you are just coming back from a particularly nasty shoot down. No wonder you are having nightmares."

"I know all that, Jonathan, but it's still hard."

They turned the corner of the No.609 Squadron Dispersal building. Several of the pilots were already lounging outside. Jonathan stopped to look into Brian's tired eyes. He whispered, "Do you want me to ask the Skipper to give you the morning to get some sleep?"

"No," Brian answered sharply, causing the outside pilots to look their way. "We are short pilots already. We can't get all of our airplanes in the air as it is. I'm needed here."

"As you were. Simmer down," Jonathan said, motioning with his hands as if he were pressing down on an overfilled suitcase. "I was just trying to help."

"Sorry, Jonathan. I know you were. I just want all this crap to go away. I just want to fly. That's all."

"So be it, then."

Squadron Leader Darling stepped outside followed by some of the other pilots. "Gather up here, lads," he said. "Who are we missing?"

"Only 'Boxer' it looks like," responded 'Waggle' Davies. "'Curly' Mansek is on break. I think he is in London."

"Correct," added Pilot Officer 'Crazy' Kradilcek. "In London."

"So it is," said Darling. "Here is the situation. Last night, a large convoy, codenamed PEEWIT, sailed from the Thames westbound through the

Channel.  They were attacked during the early morning hours near dawn by E-boats – ravaged the convoy like wolves on a flock of sheep.  Sank several and damaged others according to Group.  The *Luftwaffe* started in on them several hours later.  One One Group is on their second set of sorties already this morning."

"Bloody mornin's only just begun," mumbled 'Fog' Johnson.

Darling ignored the interruption and continued.  "Gerry has kept a nearly continuous stream of bombers and fighters after this convoy.  The convoy has apparently increased its speed to run the gauntlet as quickly as they can.  Anyway, they will be in our area later today.  We are to redeploy along with One Five Two Squadron to Warmwell, shortly.  We launch in a quarter hour.  If Gerry keeps this up, it looks like we shall have a busy day of it, chaps.  Best be on our toes.  Any questions?"  Everyone shook their heads.  "Then, grab your kit, and let's get on with it."

———

*Thursday, 8.August.1940*
*RAF Warmwell*
*Warmwell, Dorset, England*
*06:30 hours*

The process had become as routine as a morning shower and breakfast.  The flight to RAF Warmwell was as uneventful as their breakfast.  As Squadron Leader Darling walked off to the Airfield Operations building for an update on the situation and the squadron pilots settled into their normal banter and relaxation, the 'UM' Spitfires of No.152 Squadron made their approach and landing at the auxiliary airfield.  Brian wondered why the melodic sound of the Merlin engines and the sleek, curvaceous lines of the Spitfire Mark IA were still so appealing and exciting for him?  The machines continued to command his attention.  Having his own 'PR-F' Spitfire strapped to him was like a good dream come true – better than any fantasy.

Darling returned to update the pilots on the situation.  The only differences from their earlier briefing were westerly progress of the convoy . . . it was now just east of the Isle of Wight . . . and, No.11 Group was launching its third set of sorties in support of the convoy.  The stream of German bombers continued.  No.609 Squadron could expect to launch on its first combat sortie about mid-day.  A small truck brought a box of cheese sandwiches and a jug of water, which they all ate as an early lunch.  The peculiar grinding ring of the field telephone defined a probable change in their status.

They watched 'Spike' Darling nod his head numerous times without saying a word or writing any notes.  "Convoy PEEWIT is now in the Solent trying to find some protection behind the Isle of Wight.  It is now our turn. Gerry has been throwing a large number of Junkers Eight Seven's at the convoy. Most dive-bombers we've seen apparently.  They have substantial escort . . . One Oh Nines and One One Ohs.  Two Three Eight Squadron Hurris are en route.  We are to join the fight at angels one eight.  If anyone breaks off, we will return to Warmwell until this is over for the day.  Any questions?"  Again, none.  "Let's roll 'em, then."

Eleven Spitfires lifted gracefully into the air.  Darling kept their throttles up at full power and the airspeed back at their maximum climb speed, as they strained for altitude.  Brian rechecked all his switches to ensure his bird was fully armed, charged and ready for combat at any instant.  He could see other heads in the formation bobbing down to check their switches or instruments.  The black trails of smoke among the white tails of their wakes clearly marked the position of Convoy PEEWIT even through the multiple broken cloud layers.  The Germans would have no problem finding their targets.  The mêlée of aircraft above the convoy took a few more minutes to find among the clouds.  What they all saw with their 'tally-ho' call surprised and yet fascinated Brian.  The ball of fighters and dive-bombers proved too difficult to sort out properly.  With certain angles and attitudes of the combatants, the gray markings of the German fighters could not be differentiated from the greenish-browns of the British fighters.

Just as Squadron Leader Darling positioned the squadron to join the ball of fighters and dive-bombers, someone called out, "Bandits, two o'clock, level."  Brian's eyes immediately captured the distinctive swarm of twenty plus Bf109 fighters, all with red nose propeller covers.  They belonged to *Jagdgeschwader Freiherr von Richthofen Nr I*, Baron von Richthofen's namesake fighter wing.  Brian heard most of the stories, now he would match his skills against some of the best pilots in the German Air Force.

"Rooker, this is Sorbo Leader calling.  We have twenty plus fighters joining from the south.  We are engaging."

There was no response as they turned to face twice their number head-on.  The Spitfires gradually began to spread out for combat.  Brian sensed the 'PR-D' Spitfire of 'Jackstay' Beamish pulling away from him.  He followed his section leader and pushed his throttle through the frangible wire to emergency power.  The big Merlin responded with a surge of power and a deeper, more robust, throaty tone.  His right hand slid counterclockwise to the top of the circular spade at the top of his control stick so his thumb covered the red button that would soon ignite the eight machine guns in his wings.

The two groups of belligerent fighters closed rapidly as their combined closure speeds reached nearly 700 miles an hour.  Brian adjusted his flight path slightly to the left placing the pipper of his gunsight reticle on a red nose Bf109 racing directly toward him. He waited. The flashes from his opponent's wings and nose triggered him to depress his firing button. The Spitfire shuddered as all eight guns erupted. They passed in an instant.

Brian pulled back on his stick pulling the nose of his aircraft up with it. He checked his throttle full forward in emergency power before he reached up straining against the g force to use both hands to demand a little more from his machine. Aircraft filled the sky as he passed through the inverted position and rolled his fighter to skew the backside of the loop. He found an opportune target that appeared to be concentrating on one of his colleagues. As the nose came down on his target, he rolled his aircraft upright, corrected his gunsight, glanced at his slip indicator, and corrected for a slight right sideslip. As the German's cockpit began to fill his gunsight, he depressed the firing button, felt the guns respond, and then saw his adversary's masked face look up at his blazing guns. The German disintegrated in a fireball. Fragments from his victim impacted on his aircraft without immediate consequence.

Brian rolled right straining against the heavy forces to find another target. His tail was clear. The sky seemed to boil around him like a mass of carp in a frenzy over a few kernels of wheat thrown into the pond. His radio headphones filled with words from too many sources carrying too many messages, some words frantic and desperate. Brian could find nothing recognizable or useful in the fractious words. His head and eyes moved sharply and systematically, trying to gather sufficient information for him to sort through the situation around him. He reacted without thinking to the flood of events.

Targets appeared before him as he maneuvered his Spitfire to take even hasty shots at any gray-black fighter. He hit several enemy fighters without detectable result, but things happened too fast to be certain. Maybe the gun camera film would tell a different story.

He yanked and twisted his aircraft to avoid colliding with other aircraft, both friendly and enemy, as well as preventing any good shots by German pilots. The machine continued to perform everything Brian demanded of it. Just as the furball of fighters began to thin and dissipate, Brian heard the call to break off and return to base for refueling and rearming. Just as he worked to accomplish the command, a single Bf109 passed directly in front of him.

The enemy fighter trailed ugly black smoke. It was mortally wounded and probably would not make the journey back across the Channel. He rolled his opposite wing down to check for any companions or friendlies in pursuit. He found none. Brian rolled sharply back the other direction and gave chase. His 'PR-F' Spitfire Mark IA responded and gained quickly on the damaged German fighter.  He promptly found the optimum firing position and opened fire until all his ammunition was expended.  Brian followed the German for several more minutes having to throttle back to hold position.  His head and eyes worked feverishly to make sure he had no attackers.  He had to turn back.  He decided to pull along side the German.  He could not see a pilot.  Brian rolled over the top of the German and saw why.  The pilot was slumped forward either unconscious or dead.  He thought about tipping the wing of the German fighter thus sending him into a final spiral.  In the end, Pilot Officer Brian Drummond ceremonially saluted his adversary, rolling 90° to the right and pulled his machine around headed back to RAF Warmwell.

Without ammunition, Brian knew his only protection was now his instincts and skills.  He continued to work hard to avoid any engagement.  He felt the soaking sweat drenching his uniform as he dove for land.  He adjusted his flight path to steer a wide berth around the dive-bombers and fighters as the contest over Convoy PEEWIT continued unabated.  The white boiling circles in the water meant the bombing had not ceased.  Streams of black smoke in the air marked the fall of airplanes, while the burning hulks of several ships sat stationary waiting for their fatal end.

Brian was the last to land.  The squadron had returned with no losses, minor damage and numerous victories.  As the ground crews frenetically performed their duties, the pilots debriefed their first engagement of the day.  Brian would have to wait for several days before his victory was confirmed.  The excitement of boys on their first hunt floated among the pilots.  Several of them noted the change including Brian.  What had been different about this mission than all the previous sorties?  They had gone up against some of the best German fighter pilots and returned victorious.  Jokes and laughter uncommonly filled the air as they waited for the fighters to be readied for the next mission.  None of them considered this moment to be the new standard, but they certainly did want to savor the instant in time.

The field phone rang its strange call.  "Here we go, lads," said Darling, as he grabbed his flying gear.  "Mount up."

They took their leaders cue and jogged to their ready steeds.  The convoy continued moving toward them.  They were vectored directly into the flurry of ravenous Ju87 Stuka dive-bombers over the desperately zigzagging ships of Convoy PEEWIT.  Hurricanes from No.238 Squadron were disengaging, already reporting low on ammunition and fuel.  Several Bf109 fighters came down among the dive-bombers, but for the most part they were occupied by two squadrons of RAF fighters, one Spitfires and the other Hurricanes. The enemy fighters must have been very low on fuel themselves as they turned away before engaging. The dive-bombers took the hint but a little too late.

The unopposed Spitfires of No.609 Squadron swarmed over the much slower, less maneuverable and seriously out-gunned legendary German dive-bombers. Two of the attackers fell almost immediately, as Brian tried to find his target. Several times he had a good closing position only to remember that the seemingly defenseless bombers did have a rear gunner who would be frantic to find something to shoot at for defense.

The Ju87 *Sturzkampfflugzeug* airplanes quickly recognized their situation and dove for the surface of the water to protect their exposed underbelly. Brian tried several times to make a clean shot only to be greeted by the bright muzzle flashes from the rear gunner and the tracers passing all around him.

He ran the throttle up to full power, climbed and accelerated as he scanned the sky around him for possible attackers. The broken cloud cover gave him a sense of protection mixed with a strain of suspicion. What couldn't he see? Brian moved ahead of the now low-level bombers, turned and descended to face the lead aircraft head-on. He throttled back to reduce his closure speed, aligned his gunsight and let off a long, well-aimed burst of bullets. Hits flashed over the German before his target disintegrated in a cloud of parts splashing into the sea only 100 feet below. He pulled up and banked hard to avoid another Spitfire.

Before Brian could reposition for another firing run, the Germans found the protection of a low cloud layer. The No.609 Squadron pilots darted above and below the layer trying to find their targets.

"That's enough, lads. Break it off and rejoin," radioed Darling. The clouds made their join-up more difficult than usual, but several minutes later Brian counted eleven 'PR' Spitfires climbing for altitude toward the north. "Rooker, this is Sorbo Leader calling."

"Sorbo, this is Rooker, go ahead."

"We bagged three or four before they escaped in clouds. We are passing angels eight heading three six zero."

"Roger, Sorbo. Return to your staging field for turnaround. It looks like we may have more business for you."

"Persistent bastards," someone said over the radio.

Squadron Leader 'Spike' Darling leveled his flight off at 10,000 feet and adjusted his heading toward the west and RAF Warmwell. "Rooker, Sorbo. Understood."

As they flew, the white trails of the remnants of Convoy PEEWIT were now long and straight as the seamen took advantage of the brief lull. A squadron of Hurricanes patrolled overhead at about 15,000 feet. Brian carefully inspected the smooth curved lines and surfaces of his wings. The tattered and flailing remains of the red, gunport tape covers and black soot identified the machine as a shooter. He could not find any marks or damage on the upper surfaces of the wings. He knew he had several impacts from debris but could not remember any bullet hits on his fighter. A good day so far.

The hook of the Portland spit came into view through the clouds. Warmwell would be just beyond the coast. The dissipation of adrenaline compounded the gnawing fatigue from two combat sorties and nightmarish lack of sleep. The melodic tones of the Merlin even at reduced power added to the heavy weights dragging Brian into the comfort of sleep. He fought against the seduction of the numbing exhaustion.

"'Hunter!'" shouted someone.

The call caught him before he disappeared into semi-consciousness. He had drifted away to the left and the nose of his fighter had dropped toward the sea below. The several thousand feet of remaining altitude would have vanished quickly. Brian pulled his oxygen mask off and moved his goggles to his forehead. He slapped himself several times to complement the shot of adrenaline he received as he was pulled back from the edge of the abyss. He smoothly returned his Spitfire to the left wing of 'Jackstay' Beamish's 'PR-D' fighter. Several more slaps were administered as Brian felt the fatigue creeping back.

The landing was a little rough, as he bounced the machine several times, but not harmfully, just less finesse than usual. As the engine coughed to a stop, Brian closed his eyes and rested his blurry head against the stiff headrest pad behind him.

"Are you injured, sir?" came the words of a ground crewman he did not recognize.

"No, just tired," Brian responded, as he looked into the apprehensive eyes of a leading aircraftman.

"Very well, sir. Do you need any assistance?"

"No thanks, mate," he said, as he smiled trying to reassure the worried crewman.

'Jackstay' Beamish waited for him to extract himself and step to the ground. "Nearly lost you there, 'Hunter.' What happened? Are you all right?"

"Sorry, 'Jackstay.' It was just coming down from the excitement." Brian did not want to get into the poor start to the day and the accumulating, sapping fatigue that clouded his mind.

Beamish did not believe the explanation, but chose not the press his left wingman. "Are you going to be able to stay with us?"

Brian really did not want to go down that path. He pulled his shoulders back, stood up straight and took a deep, cleansing breath. "I'm OK. No problem."

"As you say then, laddie. Let's get ready for the next go, if we have one."

The intelligence debriefings had already begun when Beamish and Drummond joined the others outside the dispersal tent. The experience comparisons had already begun as they waited their turn with the debriefers.

"We are back up to Readiness, so let's look alive," Darling told them.

"The birds aren't rearmed yet, Skipper," answered 'Waggle' Davies.

Darling ignored the obvious. "Apparently, Gerry has not given up on PEEWIT. There is another tangle on at the moment, and we are next in line to relieve Two Three Four Squadron out of St. Eval."

"I guess the bloody bastards decided to get serious on us," commented 'Boxer' Stockard.

Darling looked directly to Brian. "Are you all right?"

"Sorry, Skipper. Got a little distracted."

"Right, then. We certainly do not want the RAF's newest ace to lose it on a simple return."

"Not to worry, Skipper. He has six lives left," added Davies. They laughed at the backhanded reference to his three previous losses and the fabled resilience of the cat.

"That aside, as 'Hunter's comrades, we have been woefully remiss. Our resident darkie just informed me out illustrious 'Hunter' has not only achieved five confirmed victories over the enemy, but he will be credited with two and a half more today, which brings him to nine. Nine! He is one short of being an ace twice over, and we have yet to celebrate his accomplishment."

The squadron cheered, clapped, slapped his back and punched his shoulders. They congratulated Brian as only brethren can. Brian did not feel any different other than the bone numbing fatigue. He was an ace with eight or nine victories depending on how the intelligence guys chose to record the engagements. He was now eligible for the Distinguished Flying Cross just as Malcolm Bainbridge and John Spencer had won in the Great War.

The debriefing took longer than usual since Brian had more to tell about his decision to make a head-on attack as well as the chase, which did not yield a conclusive result. They gave him ½ a victory for the second Bf109 since he had hits on the enemy fighter, and it went down. They recorded his score so far for the day at 2½ that brought his total to nine. He was now the second highest scoring pilot in No.609 Squadron. 'Crazy' picked up one more bringing his total up to ten, the highest although most of his victories were on the Continent with other air forces, and 'Jackstay' gained another ½ for a total of six.

The airplanes were reported ready although not all the debriefings were completed when the scramble order came. The pilots took off running toward their fighters. Several of the engines had already been started by the ground crews. The swirling air from the propeller carrying the gasoline fumes from the idling engine gave Brian the shot of energy he needed. Brian waited for his cue from Roger 'Jackstay' Beamish. They took off toward the south straight from their parking line as a jagged line of eleven Spitfires roaring off, retracting their landing gear as soon as the weight was off the wheels. They banked left and began to move into formation position on Squadron Leader Darling's 'PR-A' fighter.

They were given a rather low vector, which meant the fight had descended with the bombs over Convoy PEEWIT. The clouds had begun to thicken and appeared more ominous as the Sun moved farther toward the Western horizon. They barely passed 1,000 feet and the thick, black smoke columns rising from the sea identified the location of the hapless convoy caught in the horrendous air battle that continued nearly unabated for ten hours.

The third sortie proved to be the most difficult as the intensity of the day drained their energy and the paucity of slow, easy, Ju87 Stuka dive-bomber targets vanished leaving only the fighters. The continuous, aggressive, twisting, turning, climbing, diving and fleeting shots at agile targets as well as avoiding collisions and denying any attacker's shot made the third engagement seem like an eternity. Then, in an instant, the German fighters disappeared into the clouds, and the boiling sky cooled as the No.609 pilots rejoined their leader and began climbing for altitude waiting for the next wave.

At 14,000 feet, Darling entered a familiar racetrack patrol orbit overhead the significantly diminished Convoy PEEWIT. The cooler air at altitude made his sweat soaked uniform turn cold, which helped Brian fight his worsening and deepening fatigue. He could not remove his oxygen mask or goggles, with combat at altitude still possible. He pinched his neck repeatedly fighting to stay awake and as alert as he could. Fortunately, the Germans had had enough and No.609 Squadron was released to return to RAF Middle Wallop. As they crossed the coastline well into their descent, Brian pulled his mask and goggles off, again. This time he yelled several times to himself, resisting the temptation to close his eyes just for a moment. Mercifully, the return flight did not take long and was uneventful.

Brian fell asleep several times during his debriefing.  Flying Officer Royster remained patient and tolerant, and finally gave up gathering any details since Brian did not claim any hits.  He dragged himself across the A343.  He ascended the stairs to his room and fell face first into his bed where he would remain until the wake-up call the next morning.

—

*Thursday, 8.August.1940*
*No.10 Downing Street*
*Whitehall, London, England*
*20:35 hours*

**W**inston Churchill entered the Cabinet Room with Sir Edward Bridges following him and closing the doors behind them.  For the first time in nearly a week, the entire War Cabinet, less only Neville Chamberlain who was still quite ill, assembled for the evening meeting.  It had been a very long day already as a result of the extraordinary combat in the English Channel centered on the unfortunate Convoy PEEWIT.  The agenda for the evening War Cabinet meeting, which Winston had finally approved immediately before the meeting meant they would most probably be in session until midnight or later.

"Gentlemen, we have a rather lengthy agenda.  Shall we begin?  Events in and over the Channel dictate we should open with an Admiralty report."

First Lord of the Admiralty A.V. Alexander began the briefing.  "It truly was an extraordinary day and unfortunately not all together a good day.  The enemy appears to have fundamentally changed his tactics, which will most probably require us to change as well.  I shall ask the First Sea Lord to brief us on the day's naval action in the Channel and the current situation.  At the conclusion, I should like to propose a course of action."  He paused to receive a nod of concurrence from the Prime Minister.  "Sir Dudley, if you please."

First Sea Lord Admiral Sir Dudley Pound stood before the large map of the English Channel.  He took ten minutes to summarize the tragedy befallen Convoy PEEWIT in the 24-hours of its movement from the River Thames Estuary westbound through the English Channel.  The dry tones of executive synopsis did not adequately convey the consequences of the human drama played out this day.  Twenty merchant ships escorted by nine Royal Navy warships set out on their voyage the previous evening.  Three were sunk and two damaged in the pre-dawn E-boat raids.  During the day's continuous aerial bombardment, another four were sunk, six damaged most probably beyond economic repair and six more seriously damaged but still underway.  Of the 29 vessels beginning the transit, only eight did not receive substantial injury.

Sir Dudley concluded by simply restating the obvious and indicating Sir Cyril Newall would cover the air situation.

As the First Sea Lord sat back down, the First Lord cleared his throat with detectable nervousness. "A tragic day, gentlemen. The ferocity of the enemy attack on Convoy PEEWIT marks a distinct shift in enemy tactics. Combined with recent intelligence, the message seems clear. As such, the Admiralty must recommend further attempts to force the Channel with unarmed merchants should cease. While we intend to maintain a sizable warship force in the Southern approaches, we must route convoy traffic north where we will be better able to deal with offensive action."

Winston was not ready for discussion and immediately asked for the Air Ministry report. The Minister of State for Air nodded to the Chief of the Air Staff. Air Chief Marshal Sir Cyril Newall shuffled a small stack of papers. "The aircrews have already tagged the aerial action of the day as the Battle of the Stukas. We encountered an unprecedented number of dive-bombers specifically tasked with precision bombardment of Convoy PEEWIT. Double that number of both twin and single engine fighters escorted the dive-bombers. During the course of the day, more than six squadrons of our fighters were engaged in the defense of the convoy."

"Some defense," mumbled Arthur Greenwood.

Sir Cyril ignored the comment. "We are still assessing the losses for the day, so the numbers are preliminary. As of this moment, the results are best put at 30 enemy aircraft, mostly Stuka dive-bombers, shot down with approximately 20 losses of our fighters. Seven pilots were killed, and four more are missing."

"At those loss rates given the substantial difference in quantity, we may not have a viable fighter force within a month or less," observed Greenwood.

Winston wanted to interject a positive comment from the day's action but could not find one. The silence enveloping the gloom associated with the tragedy of Convoy PEEWIT had to be filled. "Are we agreed, then, to accept the Admiralty's recommendation? Is there any additional need for discussion?" Winston quickly scanned the room and was answered with somber silence. "Are we agreed, then?" He looked each person in the eyes. "Let the record show, Sir Edward, the War Cabinet and additional members of the Defense Committee were unanimous in their acceptance of the Admiralty's recommendation to route further convoy traffic north."

"As you say, Prime Minister."

"Sir Edward, if you would," Winston said, as he motioned for the War Cabinet Secretary to pass the appropriate piece of paper to the other members.  "I have taken the liberty to draft a message to Sir Hugh and Fighter Command, which I intend to transmit tomorrow morning unless there are any objections."

---

```
FROM PRIME MINISTER AND MINISTER OF DEFENCE
TO AIR OFFICER COMMANDING IN CHIEF FIGHTER
COMMAND
SUBJECT CONGRATULATIONS
BEGIN
THE WAR CABINET WOULD BE GLAD IF YOU WOULD
CONVEY TO THE FIGHTER SQUADRONS OF THE RAF
ENGAGED IN THURSDAYS BRILLIANT ACTION THEIR
ADMIRATION OF THE SKILL AND PROWESS WHICH THEY
DISPLAYED AND CONGRATULATE THEM UPON THE DEFEAT
OF AND HEAVY LOSSES INFLICTED UPON A FAR MORE
NUMEROUS ENEMY
END
```

---

"Spot on, I'd say," said Clement Attlee.

Winston scanned the room and received the concurrence he sought. The discussions of the War Cabinet continued for more than four hours as they dealt with a lengthy list of situations, issues, concerns and threats.  The U-boat offensive was becoming more focused and successful.  The need for the American destroyers even if they were second-hand, rather outdated and inadequately armed, remained paramount.  They would still be able to fill gaps in the convoy screens.  The German Navy capital ships always presented a serious threat to shipping.  It was difficult to keep track of the large, fast, heavily armed cruisers, pocket battleships and dreadnoughts.

The flood of reports from various intelligence agencies, sources and services painted a dreadful picture of massive troop redeployments toward the northern coast of occupied Europe.  The now perpetual traffic of motorized and non-motorized barges, transport vessels, riverboats and other potential invasion shipping presented an undeniable situation.  The effort of Bomber Command to slow down the invasion preparations seemed like using a garden hose on a raging house fire.  There were little detectable effects.

It was the air war that drew the most attention.  The situation at sea and to a certain extent the land forces invasion preparation did not appear to be so immediate, but the recent intelligence and the unprecedented, aggressive, air offensive demonstrated this day brought an ugly, nasty, grim specter to the war. There seemed little doubt German targeting for aerial bombardment would soon shift to land . . . factories, military facilities and maybe even cities. It was the immediacy and speed of aerial combat that bothered the leadership of the British people. The questions about the adequacy of the air defense system of Great Britain conveyed a subtle but distinctly implicit message. No one truly believed there was such a thing as an adequate defense. Virtually every member of the War Cabinet remembered Stanley Baldwin's famous premonition before Parliament in 1932 that there was no power on Earth that could protect the people from being bombed. 'The bomber will always get through,' he had said in the House of Commons. While many had laughed eight years ago, no one was laughing now. The premonition had become reality, and they all felt the singeing heat of the fire. Sir Archibald and Sir Cyril both concluded the air defense system was the best it was going to be. The key would be the fighter pilots . . . not whether they were skillful enough, but rather whether there were or would be enough of them.

The discussions, debate and decisions of the Thursday evening in early August took the time to half past midnight. Minor and routine administrative details concluded the extended War Cabinet meeting. Several ministers remained to exchange the necessary coordination words. Winston made purposeful effort to thank Clement Attlee for his support with the difficult decisions as well as to commiserate with Arthur Greenwood. Although Winston Churchill did not share Greenwood's sense of resentment, he did agree with his colleague, regarding the unfortunately stark reality the secrets shipment represented. The countless millions of pounds Sterling spent to develop those vital secrets were being simply given to a distant, rather aloof ally to minimize the impact of those same secrets falling into the hands of the tyrant.

As Churchill returned to his office, he was somewhat surprised to find Principal Private Secretary John Martin standing beside his desk with a sealed envelope in his hand.

"A private note from Mister Stephenson, sir, delivered by special courier this evening during Cabinet."

"Thank you, John. Now, you simply must get along home. The hour is far too late, and we shall most probably have another long day tomorrow." Winston actually believed they would have weeks, months and years of very long days, but that reality always seemed so negative and pessimistic to him.

"Bright and early, then, sir."

"Good night, John."

"Good night, sir," Martin said as he switched off his desk light and departed.

Winston waited until he had deposited his notebook and files in the safe and sank into his favorite stuffed lounge chair to open the envelope from William Samuel Stephenson, MC, DFC, AKA 'Little Bill' or 'Intrepid,' the head of the British Security Coordination Office in New York City, under the cover as chief of the Passport Control Office.

## FOR YOUR EYES ONLY -- PRIME MINISTER

```
DATE 8 AUGUST 1940
FROM INTREPID
TO PM
SUBJECT STATUS AMERICA
BEGIN
LAST SEVERAL DAYS VERY SUCCESSFUL BREAK
ARRANGEMENTS TO RECEIVE SPECIAL SHIPMENT
CONCLUDED BREAK EVERYTHING IS READY BREAK
WHIZZBANG TO BE MET BY SPECIAL DELEGATION FROM
OFFICE OF SCIENTIFIC DEVELOPMENT AND RESEARCH
AS WELL AS LINCOLN LABORATORY END
BEGIN
BIG BILL GREATLY IMPRESSED BY VISIT AND
RECEPTION AND HAS STRONGLY URGED OUR CASE RE
DESTROYERS BREAK SHOULD KNOW MORE WITHIN WEEK
OR TWO BREAK BIG BILL IS DOING MUCH PERSONALLY
TO COMBAT DEFEATISM ATTITUDE IN WASHINGTON BY
STATING POSITIVELY AND CONVINCINGLY THAT WE
SHALL WIN
END
```

## FOR YOUR EYES ONLY -- PRIME MINISTER

Winston reread the message several times, smiling more broadly each time. Finally, there were slivers of light in the very dark room. He thanked the good Lord above for the burgeoning relationship between 'Little Bill' Stephenson and 'Big Bill' Donovan, during the American's three week fact-finding visit last July. The two men had hit it off well, and the relationship seemed to be maturing quite well indeed. The association between those two key and influential men could very well be one of the most important in this war. How fate seemed to look after some people. He remembered the brushes with death in his own life from the campaign in India, the onslaught of the

Whirling Dervishes with Kitchener, and the bloody awful trenches of Flanders. For many years, Winston had felt mysteriously blessed. This was affirmation of those feelings.

Donovan had arrived as the unofficial emissary of newly confirmed Secretary of the Navy William Franklin 'Frank' Knox of Illinois, a prominent and influential Republican Party politician, who had been the Republican vice presidential candidate in the 1936 election. Although it had been unwritten and unspoken, Churchill believed Donovan's mission had more likely the President of the United States Franklin Roosevelt. The Prime Minister had insisted to the War Cabinet and His Majesty's Government that Donovan's mission be treated as if it had come from the President himself. Donovan had been very plain spoken about his assignment – an urgent effort to determine whether the British could withstand a German invasion attempt and remain in the fight – a prerequisite for American support.

Churchill felt a clear connection to and affinity with Colonel William Joseph Donovan – the recipient of the Medal of Honor in the Great War, a successful Wall Street lawyer, and another notable Republican Party member in service to the President. As he heard from Stephenson, people who knew Donovan often referred to him as 'Wild Bill' for reasons not entirely obvious, since Donovan presented a pronounced aire of sophistication. Churchill preferred 'Big Bill' to differentiate between Donovan and Stephenson by stature.

The Prime Minister wrote down the word 'Whizzbang' on a notepad. The codename reference would remind him in the morning to call Doctor Sir Henry Tizard, the former Chairman of the Air Ministry's Advisory Committee for Aeronautics and now Chief of the British Technical and Scientific Mission to the United States of America. He wanted to make personally certain the renowned scientist had the latest information from Stephenson and maintained the correct perspective in his detailed discussions with his American counterparts. This mission carried far more importance to the war effort than the value of the secrets they carried. This hopefully would be seen by the American President and key members of the U.S. Government as a *bona fide, gratis* offering to help further strengthen the bond between the British and Americans. The relationship carried far more weight, and Sir Henry had to be up to the task.

Winston finally went to bed at near two in the morning. He still had a smile on his face, as he drifted off the sleep.

———

*Friday, 9 August 1940*
*Headquarters, Fighter Command*
*Bentley Priory*
*Stanmore, Middlesex, England*
*14:00 hours*

The hastily called meeting by Air Commodore Hogan meant the probable receipt of late breaking information that needed to be shared with key staff members. Sir Hugh arrived just after Hogan. For Group Captain John Spencer these meetings seemed to be his salvation. The rivers of paperwork crossing his desk separated him even further from the cockpit of fighter aircraft. Being closer to the action, at least the intelligence information, made him feel just slightly closer to the pilots doing battle in the skies over the English Channel.

"What do you have?" asked Air Chief Marshal Sir Hugh Dowding to initiate the discussion.

"We received information from SIS via the Air Ministry that Göring has apparently postponed the execution of their so-called, Eagle Day. According to the message, it is the dreadfully uncooperative English weather that brought them to the decision. There may be other contributing factors such as insufficient stockpiling of ordnance to support long term, heavy bombardment operations."

"I wonder what is in store given yesterday's extraordinary attacks," mused Chief of Operations Air Marshal Geoffrey Leonard, CBE, DFC.

"Quite," responded Spencer. "Thank God for our lovely weather."

"Yes, well, the photo recce information documents the enormous build-up of both land and air forces along the French, Belgian and Dutch coasts. The onslaught is coming."

"Do you have a new projected execution date?" asked Sir Hugh.

"The 13th, at the moment."

"How assured are we this Eagle Day shenanigan is not a faint, or worse, the execution date for the invasion?" asked Leonard.

"The aerial photographic evidence has shown no rehearsal, loading or other preparatory activity for an amphibious operation. Suffice it to say, we believe Eagle Day is a significant shift in enemy air operations."

"Against cities?" asked John Spencer.

"It is possible. There is no clear means to ascertain exactly what targets they will emphasize, but it is safe to say there will be a change."

Sir Hugh cleared his throat. The small staff group turned their attention to him and waited for his words. "If I were in their shoes, I would focus on the air defense system first. Neutralize the fighters and other anti-aircraft assets, and then work on other military targets followed by industrial targets. Go after the most immediate threats."

"Therefore, we should expect concentrated attacks on our fighter forces," said Leonard.

"That would be my guess," said Dowding, "but the enemy is not always predictable."

"Ah, but the Germans have a proclivity for such predictability."

"Maybe not this time."

"Nonetheless," said Sir Hugh, "we should notify the groups."

"We shall see to that," answered Leonard. "What does the weather look like between now and the 13th?" he asked Hogan.

"For the next few days, it shall be quite inclement with low clouds and visibility, and rather steady periods of rain. About Sunday through most of next week, it should be near perfect conditions."

"Oh, brilliant," responded John Spencer. He found himself for the first time in his life hoping, wishing and praying for bad weather. Somehow the normal fall season deterioration of the weather seemed so far away, almost too far to reach.

"No wonder that fat man postponed," added Leonard. "Will they continue to harass the convoys until this damnable Eagle Day?"

"I spoke with Sir Cyril this morning. Admiralty has given up the Channel for merchant shipping. The Royal Navy shall maintain a strong presence as a deterrent to invasion, but they will route further convoy traffic north."

"Excellent," answered Leonard.

"But, then we shall be the target," suggested John Spencer. "If it must be, then let's get on with this row."

The small group laughed unexpectedly at the strident, almost juvenile challenge. Even Sir Hugh Dowding managed a slight chuckle. They commiserated for several more minutes as they watched the rain begin outside the large French doors of the commanding officer's conference room. While rain and poor weather generally brought people down, during these times, the bad weather made the Air Force officers and pilots feel a sense of relief and deliverance if even for a day or two.

———

*Friday, 9 August 1940*
*Chequers Court*
*Ellesborough, Buckinghamshire, England*
*20:30 hours*

The 16th Century, Gothic Tudor mansion at the foot of the Chiltern Hills had served as the country retreat of the prime minister since the estate was deeded to the nation in 1917.  Prime Minister Churchill had only recently decided to avail himself of the country air and scenery for even just a brief respite from the drama and intensity of the war in London.  The elegant mansion offered far more space with which to entertain guests as well as magnificent and immaculately kept grounds for a casual walkabout and quiet contemplation.  There was peace at this place, even in wartime and with the threat of impending invasion.

Prime Minister Churchill had arrived an hour earlier after the two and a half hour drive in the prime minister's 1939 black Rolls Royce Wraith Limousine with his dedicated Scotland Yard bodyguard, Detective-Inspector Walter Henry Thompson riding next to the driver.  General Ismay rode with Churchill, as he often did, acting in the role of the Prime Minister's military aide-de-camp. Churchill's personal valet Frank Sawyers travelled with the family chef Mrs. Georgina Landemare earlier in the day to allow her sufficient time to prepare a sumptuous meal for the Prime Minister and his guests.

This evening's focused dinner guest was Commander-in-Chief Middle East General Sir Archibald Percival Wavell, KCB, CMG, MC, who had arrived in London yesterday by aircraft from his headquarters in Cairo, Egypt.  The Prime Minister wanted a more personal discussion of the situation in Libya and Egypt.  Along with General Wavell were War Minister Anthony Eden, First Sea Lord Admiral Sir Dudley Pound, and Chief of the Imperial General Staff General Sir John Dill.

Mrs. Landmare accomplished her usual masterful job preparing a delight-ful meal of Scottish salmon stuffed with brown crab and bay shrimp.  Everyone raved about the exquisite supper, but they also knew why they were there.

"With our deliciously full bellies, shall we retire to the Drawing Room for cognac, cigars and more serious conversation," announced Churchill, as he stood and shuffled his guests a short distance down the hallway.  His favorite Hine cognac and *Romeo y Julieta* Cuban cigars, recently supplied by friends, were served up.  With everyone served and seated in comfortable chairs arranged in a circle in the center of the very English, high ceiling, well appointed room, Chur-chill began, "I understand the Italians decided to take the offensive yesterday."

"Yes, sir," answered General Wavell.  "I should return to Cairo as soon as our conference allows."

"Yes, certainly.  A commander should be with his troops in battle.  How-ever, before Minister Eden and General Dill release you, I thought it important to have a clear view of our situation in the desert."

"As I am sure you are well aware, Prime Minister, we are short of every-thing to conduct anything more than harassing raids on the Italian 10th Army in Libya. We have defensive lines prepared in depth, but the best we can do is contain the Italians with the forces we have."

"I am not a proponent of the defense, yet that is precisely the position we are in with this damnable impending invasion. After the dreadful losses of France and Norway, we are diminished even for the defense of the Home Islands. You must know that fact and understand it."

"I believe I do," interjected Wavell.

"Yes . . . I suspect you do. Yet, the cross you must bear to the hill hangs upon the balance between that knowledge and your actions. Let us begin with your assessment of the threat you are facing."

"The obvious element is the Italian 10th Army in Libya. General Berti recently took command. He is a very capable and experienced professional. Our intelligence services reports they have reinforced his forces with another armor division and two more infantry divisions. I would be surprised if the attacks yesterday were his main assault. I suspect they are most likely probes in force that we should be able to deal with directly. If the intelligence is correct, I believe Berti's main offensive will come in the next month or so."

"What if you take the fight to him?" Churchill asked.

"Sir, with respect, we have insufficient men, equipment and supplies for an attack into Libya."

Prime Minister Churchill knew quite well that General Wavell was correct and accurate. He also knew the necessary men and material would not be forthcoming in the short term. However, he abhorred the defense. "The best defense is a good offense, General."

"Indeed. Quite so. Yet, I must deal with reality not theory."

Churchill stared at Wavell but did not take the bait. He gestured for the general to continue.

"The Italian 10th Army is predominately infantry. While they are reinforcing their armor, they will have difficulty moving quickly. Our plan entails a flexible defense. We will bleed them as they come. We will stop them. They will not attain the western reaches. All that said, the 10th Army is not our greatest threat, just our most obvious threat." Wavell paused. Churchill bobbed his head with impatience. Sir Archibald glanced at and nodded to Sir Dudley, who remained motionless. "To be frank, Prime Minister, our greatest threat is our supply line. His Majesty's Forces Middle East are not self-sufficient. We cannot forage for our supplies. My biggest concern remains the Mediterranean. Whomever controls those sea lanes will prevail in North Africa."

Admiral Sir Dudley Pound took his cue. "We all face the same situation, General. We have a substantial force in the Mediterranean, and we expect to maintain that force. We dealt with the French last month. We are ready and looking for the opportunity to deal with the Italian Navy. We cannot protect every merchant ship. The Prime Minister has been working very hard to acquire additional destroyers from the Americans for convoy escort duty."

"Quite so," interjected Churchill. "We expect to have the agreement and those additional destroyers very soon. Sir Dudley was being delicate in his response. I will be more blunt. Those damnable German U-boats are winning the Battle of the Atlantic at the moment. We need the Americans in this fight in many more ways than one, and unfortunately that is not likely to happen for some time, at least not in the foreseeable future. President Roosevelt has promised me to provide the supplies we need, but until we gain the upper hand over Dönitz's bloody U-boats, we will be feeding the fishes and not meeting the supply needs of Great Britain. My point is, supplies will be tight for a while. We must get through the next two months, until winter weather makes a cross-Channel invasion far less likely. We should be able to keep you in basic sustenance supplies via the Red Sea and the Empire; however, armaments will be short." Wavell nodded his head in recognition. "The realities of our situation are for us, not for the troops. They must keep their morale up and they must not worry about the homeland or their families. I'm afraid you must put a bright face on a difficult situation."

"You have done a masterful job of that, Prime Minister. Your words have been the elixir the men need – truthful yet encouraging. I shall strive to be worthy."

"It is the burden we bear, I'm afraid."

Anthony Eden jumped in. "General Dill and I will do our best to find you the support you need. The day will come when we will be able to supply you properly and reinforce you for the offensive."

"When that day comes, we shall drive the Italians from Africa."

"Keep us informed," added General Dill. "Anticipate your needs as best you can. We all recognize we have a lot on our plates at the moment. There is always the Japanese nagging at our hindquarters that we must not lose sight of as well."

"I shall endeavor to make the journey to Cairo, as soon as conditions here allow. I would like to visit your headquarters, General; but, more importantly, I would like to visit your troops, to show the flag, to show them they are very much in our thoughts and prayers."

"Thank you, sir. Your visit will be most welcome."

"I shall now struggle to be a good host and avoid the details of your flexible defense plan, General. I trust General Dill has reviewed your plan and is satisfied."

"I am," answered Sir John.

"Very well, then. Let us resupply our drinks and cigars. I believe we have a movie for your entertainment and viewing pleasure. As I recall, we have American fare this evening – *'Grapes of Wrath'* with Henry Fonda – released last spring, I do believe."

The remainder of the evening was left to social intercourse with some laughter among these serious men and a serious movie. It would be after midnight for the evening to wind down. The hour and distance from London meant they would spend the night at Chequers. General Wavell planned to leave early in the morning to begin his long, multi-leg flight back to Cairo and his command. The others would not depart until later in the afternoon.

———

*Saturday, 10.August.1940*
*Thames House*
*Millbank, Westminster, London*
*10:00 hours*

Sir Henry Tizard entered the large, second floor, conference room as Big Ben marked the hour. His team was fully assembled and waiting on their leader. These were the men assigned to carry out the mission formally titled the British Technical and Scientific Exchange with the United States of America. He knew each one of them by name and reputation. Army Colonel Frederic Calvin 'Eric' Wallace distinguished himself as the commander of the anti-aircraft defenses for the Dunkirk pocket and escaped with his men on the next to last day of the evacuation; he had the most experience with the gun director problem and lately with the research and development work to produce a radio-directed fire control system. Navy Captain Hugh Webb Faulkner recently returned from naval combat action off the coast of Norway and was an accomplished naval gunnery officer. Air Force Group Captain Franklin Lawton Pearce was assigned from Coastal Command and led the first bombing raids on the German battlecruiser DKM *Scharnhorst* at anchor in a Norwegian harbor. Physicist Doctor John Douglas Cockcroft had already made several trips to the United States for collaboration with his American counterparts and would represent the British nuclear explosive research team. Edward George 'Taffy' Bowen was also a physicist by education, a disciple of Doctor Robert Alexander Watson-Watt, and a major contributor to the development of airborne variants of RAdio Detection And Ranging (RADAR) systems. Arthur Eadgar Woodward-Nutt would serve as the Mission's Secretary and was a certified aeronautical engineer with the Air Ministry. This was the central mission team. Each of them had additional supporting personnel who would assist at various levels.

"Thank you for your service and your participation in this vital mission," Sir Henry began. "Your respective ministries have assigned you as experts in your scientific or operational fields. It is my understanding that your teams have been directly involved collecting and preparing for shipment each of the items on the War Cabinet approved list for this exchange. Has everything been collected?"

Wallace answered first. "Before your assignment, we divided up the list. Each of us arranged the latest model or specimen to be crated up and quarantined, pending the final shipping details. As of this moment, all items are accounted for and prepared."

Sir Henry scanned the group. Each man nodded his concurrence with Colonel Wallace's statement.

"Excellent. I received the Prime Minister's appointment on Tuesday. Sir John brought me up to speed on your work to date. Through his ministry, we have booked passage for the entire team as well as all the exchange items aboard the Canadian Pacific ocean liner *Duchess of Richmond*, departing out of Liverpool on Friday, the 30th of August – three weeks from yesterday. You should be prepared to remain in the United States at least to the end of the year, perhaps longer. By the Prime Minister's charge, this program is far more than a simple delivery and briefing on the listed items. We are emissaries of His Majesty's Government and our country. We are expressly directed to avoid any inference of *quid pro quo* during this transfer of technical items and information."

"Then why do we call it an exchange program?" asked Bowen.

"The title evolved several months ago when the program was first envisioned and has just remained. Prime Minister Churchill was quite emphatic that this program is a transfer of technology, and more to the point, it is an insurance policy."

"Insurance against what?" Bowen asked.

Sir Henry looked down to consider his response and how much he wanted to say.

"Against invasion," Pearce responded.

"I hadn't wanted to go that far," Tizard added.

"Regardless, we are under the threat of invasion," Pearce continued. "We are currently under an invasion alert. I think the government wants to ensure the Americans at least have a head start, if the Germans are successful, and based on the last year, well . . ."

"Yes," interjected Sir Henry. "Let us focus on the positive rather than potential."

"Which is?" asked Captain Faulkner.

"As the Prime Minister explained it to me, this is as much a demonstration of our goodwill toward our American cousins. While we are not to ask for anything in return, we expect the Americans will accept this transfer and a natural, professional exchange will be generated, which is why each of you has been assigned to this team. While we will ask for nothing in return, we expect a good, thorough exchange of scientific and engineering information. At the end of the day, we share a mutual interest against a common enemy. Doctor Cockcroft can attest to the interest of the Americans."

"Quite right," Cockcroft said. "I have made three trips to the United States in the last two years. They are certainly eager to compare notes on a wide range of defense topics. I must also say they are more advanced than us in a number of areas. I suspect we will be pleasantly surprised by our welcome and discussions."

"There you have it," said Tizard. "Any other questions or statements?" He scanned the group. "Hearing none, I would like to go over the complete package for each item on the Cabinet list."

The team reviewed each item including the specific details of every physical specimen, the detailed engineering drawings associated with each article as well as the supporting technical documentation like test reports, manufacturing procedures, tooling designs, all of the associated data. With the presentation by each team member, further interaction illuminated some additional paperwork the team considered important to complement the full exchange package. After a long, thorough examination of the complete list and all of the collateral information, the team was satisfied they were ready.

"Very well, then. We are ready to go. Arthur, do you have everything you need for the record?"

"Yes, sir. I am set."

"Excellent. So, as I mentioned earlier, the team along with our equipment and documentation will depart by ship from Liverpool on Friday the 30th. I shall depart next Wednesday. I will take Group Captain Pearce with me. Our ambassador Lord Lothian has arranged lodging in Washington for all of us, as well as storage and office space. We will finalize the agenda for the specific general team meetings in Washington. My counterpart, the American team leader is Vannevar 'Van' Bush and his organization as our general sponsor – the National Defense Research Committee or NDRC. I expect we will remain in Washington, as a team, with our American counterparts, for at least a week, perhaps two weeks. After that, we will break up to sub-teams, each sub-team led by one of you . . . well, less Arthur, who will remain with me. Sir John's Home Ministry will be the supervisor support agency here, while we are in the U.S.  Arthur has completed the logistic arrangements. Arthur, if you would be so kind . . . ?"

"Yes, sir.  The Home Ministry has contracted a sufficiently sized, quarantine space in a warehouse at the Canada Branch Dock Number Two in Liverpool as well as around-the-clock armed guards for that room.  The War Office has provided escorted, military transport for the movement of your respective equipment, crates and such.  We will coordinate with each of you to move your items from their present location to the dock warehouse over the next two weeks.  We expect the *Duchess of Richmond* to dock at Number Two on Thursday the 29th and our cargo will be loaded that day.  For those who need lodging, we booked hotel rooms for you for the evenings of the 28th and 29th.  You may board the ship at anytime.  The ship will cast off at 18:00 hours sharp.  You must be aboard with all of your baggage absolutely no later than 16:00 hours Friday afternoon.  We will dock in Halifax, Nova Scotia, during the morning of September 6th.  The railway transport to move our team personnel and our cargo will be at the dock.  Once all of our cargo has been loaded and secured on the railcars, we will disembark and immediately board the special train to Washington, DC.  The only exception is Mister Bowen, who has a side trip to Ottawa scheduled for a few days, after which he will fly down to Washington to be with us for the opening meeting with the Americans.  The rest of us will arrive in Washington the following afternoon.  Lodging has been booked at the Shoreham Hotel.  We will brief everyone on the specific schedule of events in Washington, once we have linked up with Sir Henry and Group Captain Pearce.  Are there any questions?"

"Yes, I have one," said John Cockcroft.  "Most of our material and especially the documents are classified most secret.  Am I to understand you will assume protection responsibility once you pick up our crates?"

"Yes, that is precisely correct.  Once the transport team arrives at your facility, all of the cargo will be under constant armed, Home Forces guard until everything arrives at a specific warehouse at Naval Station Anacostia in Washington.  From that point, U.S. Marines, who I understand are quite diligent in the prosecution of their duties, will guard the items.  That is the moment custody will formally pass from us to the Americans.  Subsequent movement and security will be under the aegis of the Americans on specific advice from your American counterparts with your concurrence."

After a few moments silence, Sir Henry said, "If at any time you have any questions or concerns whatsoever, you will contact me or Arthur immediately.  If you are unable to reach us for some reason, you are to contact Sir John.  I know I do not need to say this, but I am obligated to do so.  This mission and everything associated with it is classified most secret.  You are not authorized to discuss any aspects or elements of the mission with anyone including your families.”

“What are we to tell our families?” asked Bowen.

“You are going to America on a special mission for His Majesty’s Government.  We will arrange for a personal telegram from you to your family to inform them of your safe arrival, but I am afraid you are not allowed to divulge anymore than that.  Your messages will be evaluated prior to transmittal.”

“Is that really necessary?” ‘Taffy’ asked.

“Yes.  This mission is too important to risk a single slip of the tongue.  I shall only note this once that in addition to the security personnel assigned to this movement, we will have several MI5 agents assigned to ensure the mission is protected.”  Sir Henry waited for that information to sink in.  “I trust each of us recognizes the enormous importance His Majesty’s Government has placed on the success of this mission.  Let us prove worthy.”

The meeting ended.  The team members departed and would not rejoin until Liverpool, and the entire team would not reassemble until they were all in Washington.

———

*Saturday, 10.August.1940*
*RAF Middle Wallop*
*Middle Wallop, Hampshire, England*
*16:15 hours*

The rain continued to come as a heavy mist throughout the day.  For some odd reason, the No.609 Squadron pilots remained at Available status all day except for a fall back to 30-Minute Available to allow the pilots a hot meal for lunch at the Officer’s Mess.  The only resident squadron not at some alert status was No.604 Squadron, the night fighters with their hulky twin engine Blenheims.  The night pilots were still asleep, or at least most of them were.

It was not until late afternoon that their usual discussion turned in a different direction.  For Pilot Officer Brian Drummond, the mercy of his brethren, whether intentional or not, kept his mind off the personal significance of the day.  He tried not the think about it, and the pilots kept a steady stream of jokes, levity and frivolity as they tried to occupy the agonizing waiting time.  It was ‘Fog’ Johnson that finally broke the flow.

“Why don’t they just let us go, for God’s sake,” said Flying Sergeant Miles ‘Fog’ Johnson.  “We’ve been sitting here all day in this incessant rain.”

"Your wally has a bit of an itch, does it now, 'Fog?'" jabbed 'Jackstay' Beamish, producing a robust peel of laughter. The youngest of the pilots had a reputation with the ladies almost as renown as the late Flying Officer Reginald 'Organ' Foxworth.

"You could say that. I would certainly be doing something more productive."

"Literally, I should think," added 'Waggle' Davies.

"Not my fault you old sods can't stand up to the line."

The challenge evoked the expected response as the group began a light-hearted series of jabs at their choices in women. While everyone left Brian out of the jocularity, he tried to force himself to think of Rosemary Kensington, or Mary Spencer, although her image brought strings of guilt, and even an occasional thought of Rebecca Seward, his former girlfriend from Kansas. The reminiscences of the experiences with Anne Booth proved impossible to avoid. Although he had known her a year, she had a significant impact on his young life. He admitted to himself he loved her more than the other women. The torrent of questions flooded back with every memory.

Brian took every opportunity during the brief breaks in the rain to walk down the flight line among the Hurricanes and Spitfires poised for combat. The airplanes always made him feel better about things. He walked through the maintenance hangar, found Gordon, Toldson and Jenkins, and talked about the machines and plans for high tempo operations. Questions about the threatened invasion generally filled most conversations. Their spirits remained high, and although the future looked rather dark, everyone in or out of the military seemed to find good things to interject among the questions.

After his third walking tour, 'Jackstay' Beamish waited for him outside the Dispersal building even though the light rain had returned. "Lord Morrison called, Brian. He wants you to call him back. I would suggest you use the telephone in the visitor's room at the Operations building."

"Thanks."

Brian knew what the subject of the telephone call was probably about. He did not want to hear Jeremy's voice and especially his news quite yet, but he knew he had to get this behind him. He took a small notebook from his trouser pocket, leafed to the correct page and found Jeremy's number at No.74 Squadron at RAF Hornchurch.

The courteous words of the squadron clerk answered the call. "This is Pilot Officer Drummond returning Flight Lieutenant Morrison's call."

"One moment, sir," the man said.

Brian thought to himself, he was glad they had Corporal Warren. Having a pleasant female voice on the telephone and announcing combat scramble launch commands as well as the filtering effect of her presence made Brian thankful for her contribution to the squadron.

"Brian, this is Jeremy."

"How are you?"

"Not good, I'm afraid.  Give me your number, and I'll call you right back."

He had to look at the card beside the telephone.  "Middle Wallop two five three seven."

"Got it.  Call you back," he said, as the line went dead without a response.

The return call took less than a minute although it seemed much longer. Jeremy was out of breath as they dispensed with the usual amenities. "Brian, you know why I am calling.  I have used my connections to keep track of things for us.  Virginia and Anne as well as three of the others were hanged this afternoon at four o'clock.  They are gone, Brian, and I am so terribly sorry."

The withheld emotions and the isolation of the empty room brought a stream of unchecked tears to Brian's face. Jeremy let the silence span the gap.  Brian swallowed hard several times covering the mouthpiece of the handset.  He fought to prevent the image of her death entering his thoughts. He wanted only the subtle sweet smell of her light brown, curly hair, the soft, smooth curves of her body, and the delightful tone of her words to complement the image of her pretty face.

"Are you all right, Brian?" Jeremy finally asked.

Brian cleared his throat and tried to regain his façade.  "Yeah, thanks, Jeremy.  It's turned out to be harder than I thought.  I loved her, Jeremy," he said, as the tears returned.

"It is OK, Brian.  It is OK to love her.  She was a good person and a beautiful woman."

"You've lost Virginia.  I know you must feel the same."

"Yes, I do.  I have dealt with my grief as I usual do, drunkard old sod that I am."

Brian laughed, which was good.  He could see the antics of Flight Lieutenant Lord Jeremy 'Mud' Morrison, Esq, as he had seen numerous times before. Jeremy had always been a humorous and pleasant drunk. Brian thought for the first time about why Jeremy felt the need to call him with the news.

"Jeremy, I need to tell you it's all right as well. I'll be forever grateful you introduced us despite what has happened."

"Jesus, Brian," he answered, with his own choked back tears. "I am so terribly sorry." His voice trembled. "I never dreamed it would turn out this way. I only wanted the best for you, our glorious Yankee volunteer."

"Thanks, Jeremy. We did have the best. We didn't do what they did. They chose to do it. We simply loved them and enjoyed life with them."

"You are wise beyond your years, Master Brian."

"I don't know about that," Brian said, feeling a little embarrassed. He did not say the words as wisdom . . . only what he felt.

"Yes, well, seems like an appropriate day for such a tragic event."

"My feelings precisely."

"It usually clears up for you first in the West. How is it?"

"Still raining. We are being held at Available."

"We as well. All bloody day. A dreadful way to spend the day."

"Better than facing three or four times your numbers in Hun fighters."

They laughed. "Just think of it this way, 'Hunter.' We do not have to worry about finding a target to shoot at up there. Everywhere you turn, you can find a good Hun bastard."

Jonathan Kensington knocked on the door and poked his head through the small crack. He held up his left thumb as an inquiry. Brian held up his right thumb as an instinctive response although everything was not OK. He whispered at a level barely audible, "We have been brought to Readiness." The words meant the skipper wanted him back at Dispersal in case they received a launch command. Brian nodded his head, and Jonathan closed the door.

"We've been brought to Readiness," Brian repeated for Jeremy.

"Go get 'em, tiger. When might you be in London next?"

"I don't know," Brian responded. "I'm suppose to get my break on Monday, but at the rate things are going I probably won't get it. I'll call you the next chance we get, and maybe we can meet at Shepherd's."

"Excellent. I'll look forward to it, Jeremy, and thanks for thinkin' of me."

"No problem, kid. I am always thinking of you. Keep the ball centered," he said, as he hung up the telephone.

The fighter pilot's admonition to one another to keep the aircraft trimmed without sideslip for a good clean firing position brought a warm smile to Brian's face. He hung up his handset and left the room. The rain had broken, and the skies looked like they were beginning to clear. He looked at his watch.  It was 18:17.  They had been waiting for eleven hours.  Maybe they would get a simple, quiet patrol to at least shake out the cobwebs.

The other pilots left him alone when normally there would be jabs at getting a social telephone call during alert hours.  They knew what news Brian had received, and there was no need to rub salt, however unintentional, in an open wound.  He was thankful for the mercy as he tried to regain his fighting face, as they called it.

The squadron stood at Readiness for another three quarters of an hour before they were released for the night.  They hurried across the facility to catch the last serving of the evening meal.  Several of the pilots had private nocturnal activities planned, while most retired to the bar for beer and spirits.  In an effortless, easy going, manner, the pilots shared flying stories to divert Brian's thoughts from his loss.  Again, mercifully, there were no tales of female conquests that might antagonize the wounds.  He also resisted the temptation to drink to a level of knee walking drunkenness.  His brethren did their part, and Brian was quietly most appreciative.  The evening passed quickly.  Brian went to bed early.

—

# Chapter 3

There aren't any great men.
There are just great challenges that ordinary men
like you and me are forced by circumstances to meet.

-- William F. Halsey

*Sunday, 11.August.1940*
*RAF Warmwell*
*Warmwell, Dorset, England*

## Week 6

They knew this was not going to be an easy Sunday when awakened not long after dawn and redeployed to RAF Warmwell before breakfast. They were now back up to twelve aircraft although several of the birds were beginning to show the wear and tear of extended combat. The clouds had cleared by the time they arrived at the coastal staging field. It looked as if it was going to be a good weather day that meant they could expect the Germans in force since activity had been almost non-existent during the last two days of poor weather.

Breakfast matched the rather austere conditions at Warmwell . . . cheese sandwiches and hot tea. It gave the pilots and ground crews something to focus their attention on instead of the corrosive effects of waiting for someone to attack. They barely finished their tea when the order to patrol over a small forming convoy in Portsmouth harbor came into the tent. All twelve Spitfires were in the air in good form. They arrived overhead the famous port city at 15,000 feet and entered their loiter pattern to wait for an enemy not yet identified. They waited for 30 minutes before the radio call came.

"Sorbo Leader, this is Bandy calling."

"Go ahead, Bandy."

"Sorbo, possible twenty plus bandits approaching from the southeast. Climb to angels two zero and turn to heading one three five for intercept."

"Understood, Bandy," Darling radioed, as he turned the flight and began the requested climb.

The carpet of puffball clouds like polka-dots over the green sea lay out far below them with clear, unmarred, blue sky around them. These were perfect conditions for the fighters – no place to hide. In the morning sun, their targets would be on the right side so they could dive on them from out of the Sun. All the switches were in their proper position and the instruments reported the fighter's vital signs in excellent health, the bird of prey was ready for the kill.

"Sorbo, this is Bandy calling. Turn to one seven five."

"Understood, Bandy."

This would be the final course adjustment for their perch position, unless the incoming raid changed course.  Brian could not see the raiders.  None of the others reported, tally-ho, which meant they did not see the target yet either.  They waited an inordinately long time heading generally south toward occupied France at 300 miles an hour.

"Sorbo Leader, this is Bandy calling.  Raid appears to have turned around and too far to give chase."

"Understood.  What next?"

The pause told them the controllers were thinking and conferring among themselves since this was not a usual occurrence.  Darling throttled back first and turned the flight back north.

"Sorbo Leader, this is Bandy calling.  Turn to heading three two zero and return to patrol station."

"Understood."

Brian imagined they were all laughing as he was in his oxygen mask.  Sometimes the ability of the controller to 'see' into the cockpit was extraordinary and mystical, and other times, their inability to relate to what was going on in the air seemed baffling, disappointing, and as in this case, humorous.

Darling kept their power back to maximize their endurance.  The return flight took considerably longer at their endurance speed around 120 miles per hour.  By the time they reached their patrol position over Portsmouth harbor, they had been airborne for an hour.  If a raid developed, they would have much less time to deal with the enemy fighters, let alone go after the bombers.  Brian found himself becoming impatient about fuel.  They needed an airborne fuel bowser during situations like this, but they had not found a means to perform the trick.  As the Merlins continued to burn fuel, the usefulness of the No.609 Squadron Spitfires dwindled rapidly with their fuel gauges.  The fighters needed fuel to carry out any effective combat.  While the Bf109s only had about 30 minutes of effective combat time over Southern England, the Spitfires needed at least that much time to force the Germans to break off the engagement giving the RAF a brief moment of advantage.

"Sorbo Leader, this is Bandy calling.  We have another inbound raid.  Return to your forward station for rapid refueling.  We have Maida airborne to relieve you."

"Bandy, this is Maida Leader calling.  We are climbing through angels two, heading one six zero."

"Roger, Maida.  Continue on heading climb to angels one seven."

"Understood."

"Raid appears to be fifty plus bandits at angels one five."

"Understood."

"Go get 'em, 'Two Decker.'"

"Thanks, 'Spike.'"

They listened to the radio directions from Middle Wallop Sector Control to No.152 Squadron. As they approached the coast, they saw the distinctive spread of V formation Spitfires climbing from below them and off to the left of their track. They landed back at RAF Warmwell with their gunport tapes still intact.

Darling gathered the pilots near the middle of the line of Spitfires as the ground crews moved like warrior ants over the fighters, refueling and checking each machine. The pilots took the moment to stretch their muscles as they always did after a mission.

"Gerry was probably trying to time his approach, draw us up, and then catch us with low fuel."

"Nasty bastards," responded 'Waggle' Davies.

"Surely an indication they have felt the sting and are trying to find ways to minimize their pain."

"They have not felt the worst of it, yet," added 'Sparky' Morrow.

"I shall make a quick call to Sector to see if they still want us," said Darling, and then walked to the Dispersal tent.

The discussion of German tactics continued as the ground crews reported the fighters ready for flight. Johnson, Komer and Koenig sat on the grass, and then lay down to take a nap, however long they might squeak out. Just as Brian decided the three recumbent pilots were probably the smart ones, Darling returned with the launch order. They were not in a hurry, but they moved through the procedures smartly. They found the battle already joined off the coast near Portland Spit.

An as yet unknown Hurricane squadron occupied the dive-bombers low, quite short from their target, while No.152 Squadron ground away at the defensive circle of a squadron plus of Bf110 *Zerstörer*, twin-engine, twin-tail fighters. Brian still found the tactics humorous. The scene of twenty rather large fighters following one another around a large circle and a swarm of Spitfires diving, slashing and cutting through the circle was too clearly defined to avoid the comparison to the old West. Most of the Spitfire pilots, while they had respect for the guns of the Destroyer fighters, found the German tactics quite to their liking and usually made fairly quick work of it.

"Bandy, this is Sorbo Leader calling. We have a tally-ho on Maida as well as the bandits. We are rolling to join," Darling radioed, as he rolled the squadron to join the fight.

"Sorbo, Maida.  Welcome to our party.  We are low on petrol.  We'll break off and let your blokes have at these dumb bastards."

"Thanks, Maida."

"No problem.  Good hunting.  A bit like ducks on a pond, I'm afraid."

As they approached, the No.152 Squadron Spitfires turned sharply and dove for the sea to disengage.  Several of the German twin-engine fighters thought they had the upper hand and rolled out of their defensive circle to pursue the fleeing Spitfires.  It proved to be a fatal error in judgment.

"'Sparky,' you take the high ones.  I'll take 'A' Flight after the low ones."

"Understood."

Darling led Blue and Green Sections after the seven Bf110s trying to catch the Maida Spitfires.  "Take your own," was the simple call that meant they were to pick out their targets among the diving Bf110s.

Brian waited for 'Jackstay' Beamish to pick out his target.  He found a lone Bf110 well to the left of the others and pushed his throttle up through the emergency gate.  The engine and fighter responded well to the demand.  His sight picture was near perfect. Brian scanned quickly all around him to see if he was being chased.  He was not.  As he bore down on his target, a single flash from the rear seat gunner indicated his discovery.  As Brian approached his firing point and began to press on his firing button, he saw the gunner frantically working on his jammed gun.  The German pilot recognized his predicament too late as Brian opened fire with a precise burst that laced the Bf110 Destroyer.  Parts of the wing flew off passing just beneath his Spitfire.  As Brian jockeyed his fighter for the *coup de grâce*, the outer half of his target's left wing buckled, snapped back over the fuselage and tore loose causing the Bf110 to roll over onto its back in a final spiral to the sea.

Brian pulled the nose up to give himself a good look around him.  'B' Flight continued to work on the remnants of the Bf110 defensive circle about 3,000 feet above him.  The 'A' Flight efforts against the breakout German fighters appeared to be more successful.

He could see only four Bf110 fighters with five of the 'A' Flight Spitfires giving chase to the diving targets trying to escape.  Brian rolled onto his back and pulled the nose toward the sea below.  He had the angle on the engagement and maneuvered to close on the group of Germans.

As the scream of high speed air rushing past his canopy reached deafening levels, Brian saw the Germans would reach wave top level before he could close the distance.  The chasing Spitfires were outside firing range and did not appear to be making up the distance quickly.  Brian adjusted his flight path to flatten out his dive and give himself a reasonable position for a slashing shot. The Germans saw him coming and opened fire early. The red balls rose up to him, and then fell right and behind him. The gunners were having difficulty compensating for the angular rates and the closing speeds. He judged his pull up point to flatten his dive before he reached a proper firing point on the closest German fighter. The red balls moved faster and closer to his fighter. He thought the Germans would win the contest. He opened fire early using an aim point well ahead of his target.

The stream of bullets bent to the right behind his target. He squeezed off one more burst, and then pulled up sharply to avoid being hit by the rear gunners. He rolled right to see his brethren closing at various angles, and then rolled sharply back to the left. Several of the Spitfires passed beneath him as he maneuvered to regain a good angle. All six Spitfires chased the remaining three Bf110 fighters until it became obvious it would take too much time and distance to catch the fast, twin engine fighters. A large cloud of smoke, spray & boiling water marked the final moment of Brian's target.

"Let's break it off," radioed Darling, as he pulled up gently and throttled back below the broken emergency seals.

Squadron Leader Darling listened to the limited radio traffic until he was satisfied they had done their job in chasing off the enemy raid. "Sorbo Red, we are southwest and low."

"Sorbo Leader, tally-ho. We'll join on you."

Darling waited to gain a little more altitude before he called Sector Control. They were cleared back to RAF Warmwell. The radio turned silent in what would probably be a lull in the action allowed the committed squadron's to return for refueling, rearming and a rest for however long they might be permitted. They did not shut their engines down until 14:30 hours, well after lunch. The station staff at virtually every RAF facility kept fairly strict hours on the Mess, even the makeshift facilities at Warmwell. Darling managed to scrounge up some cookies and tea that proved adequate to cut the hunger pangs they all felt. The missed meals in the past never seemed to bother the pilots, but as the tempo of operations increased and the intensity of the air war picked up, their tolerance for such nuisances diminished rapidly.

The squadron remained at Readiness through the late afternoon. Rumors began to make their way to Warmwell, as they always did during quiet times. Brian always felt it was a natural process of occupying their minds as they waited for combat to come to them. No.11 Group covering the Southeast was still in action over the eastern portion of the Channel. The word spread quickly that the Germans had intentionally bombed land targets. They attacked the Chain Home stations in Kent and several of the anti-aircraft batteries protecting Dover, Falmouth and Brighton. If true, the war had definitely taken a serious turn. None of them said it aloud, but Brian was certain they all thought it . . . Southampton, Portsmouth, Bournemouth and Plymouth would soon be next.

They came to Standby with news of a large raid heading toward Southampton. Squadrons from Tangmere and No.11 Group would be tasked with the response. No.609 Squadron was brought to full alert readiness in case they needed to help. They waited for the longest time strapped into their fighters expecting the launch command at any moment. After two missions by early afternoon and the remainder of the day at Readiness or Standby, patience was wearing thin. They wanted to go.

'Crazy' Kradilcek reached his limit. He unstrapped and stood up on his seat to stretch his muscles and move his joints. Several pilots were laughing.

"Hey, 'Crazy,' do like your fancy slippers," yelled 'Sparky' Morrow from the Spitfire off Brian's right wing.

Brian felt the urge to do the same as the young aircraftman who served as his crew chief this afternoon asked about flying combat and having to stay in the small cockpit for so long. The distraction helped Brian forget the tension.

"You are beginning to perspire, sir," the young man said, as he looked over to 'Crazy' still standing on his cockpit seat.

His concession came in the form of an umbrella held by the considerate young man. It did feel better without the sun bearing down on him. Sitting, strapped to his seat with the Vickers-Supermarine Spitfire Mark IA poised to take flight in defense of Great Britain, the air stopped and pooled around him like a stale, tiny room. The umbrella helped, but the stagnant air choked the breath from him. The propeller wash would feel better even with the unburned gasoline fumes from the Merlin's massive exhaust manifold. He pulled off his helmet. His hair was soaking wet, which added just a sliver of relief, but not enough to prevent frustration from taking root. He wanted to scream to Squadron Leader Darling across the four intervening airplanes . . . were they going to launch or not?

Just then, he heard the first Merlin fire off, followed by several others. Brian moved quickly, as his frustration vanished in an instant. 'Crazy' Kradilcek dropped into his cockpit like the bottom had fallen from underneath him. Brian cranked the engine, which came to life promptly, as usual, especially since it was still warm from the earlier flights and the afternoon sun.  Missed the radio calls, but the routine was always the same.  He would follow 'Jackstay' Beamish on the other side of 'Crazy,' and he would be able to catch up with the check-in call once they were airborne.

No.41 Squadron had been in a tangle with a raid on Southampton.  The bombers, this time apparently Ju88 medium bombers, had dropped their explosive loads on the harbor and departed.  A second band of Bf109s had arrived to enable the first escort group to disengage.  Their job was now to help No.41 Squadron disengage, since they were running low on fuel and ammunition.

They jumped into the fight with a vengeance.  The white and yellow nose cones meant at least two squadrons of Bf109E-4 fighters, and they were good.  The tangle of fighters and the very brief shot opportunities kept his head on a constant swivel while pulling his machine hard.  His uniform was soaking wet.  His neck burned from the abuse, and his arms ached from the heavy control forces needed to maneuver his Spitfire.  This was not the same fight they experienced earlier.  Things happened in terms of heartbeats, not seconds or moments.  The twisting, turning fight ended like someone rang a bell at a large, one-size-fits-all, boxing match.  The No.609 Squadron Spitfires gave chase, but the near speed parity left little energy to close and the Bf109s were lighter with less fuel and ammunition.

Sector Control sent them back to RAF Middle Wallop for probable release although that was not said.  By the time they arrived back at their home base, the stories about the day's events came faster than could be absorbed.

"We had a big day," said 'Sparky' Morrow, but it sounded like everyone else had a bigger day

"My blokes say, 'Sailor' Malan got his 11$^{th}$ kill today," added 'Fog' Johnson.

"Everyone's talking about it.  He's the highest now."

"Wonder if Bader or 'Bobby' Tuck will catch him?"

"One of the lads from Six Oh Four says they had a big ta'do this afternoon while we were tangled up with the One Oh Nines," said 'Jackstay' Beamish.

"Heard about it."

"According to 'Tiny' Tinker, One Five Two lost one and had several in the water.  They launched Two Three Eight along with several Blenheims from Six Oh Four to search for the blokes in the water.  Found the lads for the boats, and then found a Heinkel Five Nine with a buzz of One Oh Nines.  The Hurris took on the fighters while the Blenheims chewed up the Heinkel."

"I still don't like this idea of shooting down a float plane painted white with big red crosses all over it," added 'Red' Burns.

"Bastards deserve to die," said 'Crazy' Kradilcek in his adequate English with considerable fervor.

For some reason, the vehemence struck the pilots as particularly funny touching off several waves of rolling laughter. Brian knew it was serious business, but the way Kradilcek spit out words just seemed funny. They all felt some resistance to going after the unarmed He59 rescue planes, but the order and explanation from the Air Ministry was quite clear. They rationalized the orders against the intelligence that told him the floatplanes were being used for intelligence gathering and targeting as well as rescue work. No one ever tried to portray war as glamorous or pretty.

Darling joined them while they were still laughing, some to the point of tears. "What's so funny?" he asked. Several of them tried to explain without success, driving them deeper into uncontrolled laughter and incoherent speech. "Fine. We are released, gentlemen. It would seem we have plenty to talk about in the Mess."

They did indeed have much to discuss. This day was the single most successful day for the squadron in its history. Nine enemy aircraft shot down. More than double that number damaged to various degrees. They had some damaged aircraft, but everyone returned without serious injury to body or airplane. Pilot Officer Brian Drummond was credited with his 10th victory, moving into a rather rarefied group. Squadron Leader 'Sailor' Malan, Commanding Officer of No.74 Squadron, led Fighter Command and the RAF as the highest ace so far. They had plenty to celebrate, but none of them could ignore the marked change in the air war over Great Britain.

———

*Monday, 12.August.1940*
*RAF Middle Wallop*
*Middle Wallop, Hampshire, England*
*08:00 hours*

The day began early in a herky-jerky sort of way from their wake-up call through an unusually disjointed breakfast and several runs up the alert order, even to Standby in the cockpit once. The pilots grumbled and grunted, but more to fill space than to complain about something that they knew was uncontrollable. For all the pilots including Pilot Officer Brian Drummond, the alerts took their toll. No matter how hard they tried to ignore the tension of approaching combat, they still suffered the effects of adrenaline rush and withdrawal. In a peculiar way, Brian enjoyed the measured shots of adrenaline, and the energy and focus it brought with it, but each episode seemed to use capacity, as if the human body could only tolerate so much.

Brian and Jonathan often talked about the feelings they had at different times in the conduct of their duties. They often wondered why the reality of death's proximity did not seem to take the edge of the charge they received from the challenge. With the rapidly growing intensity of the air war coupled with the near perfect, clear, but slightly hazy summer day, this was going to be one of those days that would severely tax what enjoyment they derived from this endeavor.

The telephone ring brought them all to attention like Pavlovian dogs. "Scramble," announced Squadron Leader Darling.

The pilots responded instantly as those still wearing their flight gear bolted for the airplanes and the others scurried to grab their flying kit. Engines began to start up before they reached their fighters. For Brian, Leading Aircraftman Bernie Gordon had the 'PR-F' Spitfire's Merlin III engine purring at fast idle as he stood on the left wing root in front of the cockpit with his back to the big propeller to shield Brian from the propwash as he jumped into the cockpit. As Brian plugged in his radio and oxygen fittings, Bernie placed Brian's shoulder straps over him and helped finish buckling him into the seat. A quick scan of the cockpit produced a thumbs up to the crewchief. Brian waited for several seconds until 'Jackstay' Beamish's 'PR-D' Spitfire began to move, and then looked to Bernie holding the wheel chocks and saluting him. Brian returned the salute and advanced the throttle slightly to begin his movement.

The line of Spitfires taxied fast to the crown of the slight hill that distinguished the aerodrome, and followed 'Spike' Darling as he pushed his throttle to full power and raced across the large grass field toward the south. The whole of No.152 Squadron was also in a scramble event along the western edge of the airbase to their right. They would be just behind the twelve No.609 Squadron fighters. When the wheels broke ground, Brian raised the landing gear and felt the machine surge forward still skimming the grass. They picked up speed swiftly, as they cleared the treeline on the south side of the airfield and began drifting toward their combat positions.

Brian felt the energetic pumping of his heart as he quickly armed his fighter, to be ready for instant combat, if required. Both Sorbo and Maida checked in with Sector Control, and received their instructions. Climb to 20,000 feet. Two Spitfire squadrons to that altitude meant a large raid with many fighters. Many eyes searched the sky although there were no reports or indications the enemy was close to them. Eventually, they heard the call. Although it was not stated over the radio, Brian recognized that the Chain Home

RDF network had picked up a 100+ size raid.  It was the first time any of them had heard that number in describing the size of an inbound enemy attack.

"This ought to be interesting," someone said over the radio.

No.609 Squadron as well as No.152 Squadron were tasked with engaging the two principal layers of fighters above the largest formation of He111, Do17 and Ju88 medium bombers any of them had seen.  Brian remembered, as probably others did as well, the early days of the war while they were based at RAF Drem in Scotland, when they saw three or five bombers and thought that was a lot.

No.152 Squadron positioned to take on the lower layer of both Bf109 and Bf110 fighters, and No.609 Squadron moved to engage the higher layer of only Bf109 fighters.  Brian noticed at least two flights of Hurricanes closing with the bombers.

As the two groups began to sort themselves out, the top layer of fighters turned directly into the Sorbo fighters.  Dark puffs of smoke dotted the eastern shoreline of the Isle of Wight.

"They're after the Chain Home station at Ventnor," someone radioed.

"They're trying to poke our eyes out," someone else said.

"Shut your gobs, lads," Darling commanded, "and let's go stop 'em."

The sky exploded, as airplanes twisted together in three dimensions.  The fighters made far more difficult targets than the plodding Ju87 Stuka dive-bombers or the less maneuverable medium bombers.  Shot opportunities flashed before them.  The reflexes, concentration and energy demanded of each pilot reached the limits of human performance.  The calls and sometimes shouts of warning and fear added to the confusion in the skies over the Isle of Wight.  The battle raged for what seemed like an eternity, and then calls of no ammunition began to join the cacophony.  Brian reached the same point along with his mates.  Fortunately for them, the Germans ran low on fuel, or ammunition, or both, and the battle ended as it had begun.

The return to Middle Wallop allowed a careful assessment of his machine, or at least as much as he could see.  He expended 2,400 rounds of 0.303 caliber ammunition through his eight Browning, air cooled, machine guns, and to the best of his knowledge did not hit anything nor was he hit by anyone else.  His sweat soaked uniform, lack of ammunition and only a quarter tank of fuel made him feel as if he had done a lot, but he had nothing to show for the effort.

As they approached RAF Middle Wallop, they all noticed the columns of smoke rising from the airfield.  The Germans had gotten through the defenses and bombed their home base.  They approached for landing to the west, uphill. Brian's wheels touched down giving him the characteristic rumble beneath him. All of sudden, a bomb crater appeared directly in front of him. He had aircraft on either side. Brian jammed the throttle up to the stop. The Merlin responded to command. The aircraft accelerated quickly, but not fast enough for his closure with the crater. He pulled back on the stick even though he did not have flying speed. The Spitfire lifted into the air briefly, and then stalled and promptly returned to the grass. It was just enough to pass over the hole in the ground. Brian allowed the fighter to accelerate again. The nose felt very heavy as the Spitfire took to the air. He remembered he still had the landing flaps down. Brian quickly raised the flaps. He looked both directions to see if anyone had the same problem he did. One Spitfire was to his right and slightly behind him. None to the left. Brian banked his machine left, leveled his wings once established on the downwind leg, and then pulled the throttle back to level off 500 feet above the airfield and to hold his airspeed below the limit with his wheels down. As he moved downwind, he looked out at the landing area. Maybe a half dozen craters pocked the landing area toward the north side. Why hadn't he seen those craters prior to his initial landing attempt? Why hadn't the Tower warned them? He found a clear landing line and safely landed his 'PR-F' Spitfire. He taxied between several other craters before he parked in his place in the line and shut down the engine.

"Bit of a problem with the bomb craters, ay, sir?" asked Leading Aircraftman Gordon.

"Nearly fell into the friggin' hole, Bernie. Guess we got bombed while we were out."

"Three Dornier One Sevens dropped some presents to us. No major damage so we hear."

Brian took a good look at his bird. His in-flight initial assessment proved correct – no damage. "She looks good, Bernie. Let's get her ready."

"Aye, sir, in a flick."

Brian walked to the Dispersal building. Half the pilots waited in the warmth of the Sun, while the remainder were inside being interviewed by the intelligence folks.

"Thought you were going to try to bury that thing," joked 'Sparky' Morrow who landed on Brian's left.

"Damn crater surprised me."

"You are supposed to look before you land there, sonny," offered 'Red' Burns, as part of the friendly gibes.

Brian knew this would only continue until someone changed the subject. "I guess we have proof the Germans are doing something different," he said.

"Tad bit of an understatement, I'd say laddie," said 'Jackstay' Beamish.

"The Skipper was tellin' us that Gerry hit the Southeast pretty hard," said Burns.  "They've hit nearly every airfield in the southeast of England from what we hear."

"Hit Manston, Kenley and Biggin Hill rather hard, so they say," added 'Boxer' Stockard.

"And, just about every Chain Home station along the southern coast," 'Waggle' Davies said.

"At least there is no need for speculation anymore.  We know the Gerry bastards are serious," said 'Sparky' Morrow.  "We've our work defined rather well, I'd say."

Most of them laughed as they had done many times, playing dumb as if they could not figure out what the Germans were going to do.  They liked to joke about wondering when they might get some clues regarding German intentions.  The two Czech pilots smiled and tried to laugh, but often did not understand the humor in a subject so serious and personal.  'Slim' Koenig did not look well and had not looked well for many days although he insisted he was fine.  Brian and Jonathan felt it might be nerves like 'Hank' Maxwell or 'Angle' Ashcroft.

The intelligence debriefing ground on as they ate simple meat and cheese sandwiches, and tea brought by the mess stewards.  They would find no relief from alert duty as they had in the past.  With the invasion alerts, questions from the civilian population they came in contact with and the obvious change in German tactics, the sportsman's class of gentlemanly distant war had finally come home in the most intimate level.

As they continued to joke about the war around them, Brian thought of Charlotte Palmer.  He wanted to know her better before the war and the invasion consumed her or displaced her.  He felt an unusual bond to her as if she had given him life, not like a mother, but in some other mystical, heavenly manner.  He tried desperately to not think of Anne.  He did not want to forget her.  He just could not afford to think about her.  If he did, images of what came between them inevitably crept into his thoughts.  Brian still kept his thoughts of Rosemary Kensington and the possibility of being closer to Jonathan.  He actually considered marriage, to become part of the family, someday, but certainly not under these conditions.  Even Mary Spencer, despite the illicit and enormously risky affair as well as the potentially disastrous consequences of their relationship, gave Brian many warm thoughts.  But, it was Charlotte Palmer that captured his quiet attention.

The telephone rang. Brian and several others caught the panic in 'Slim' Koenig's eyes before he ran around the corner of the building to vomit. Those that had seen the expression and result looked at one another but did not speak. They knew the intensity of the last few days of air combat probably took a terrible toll on 'Slim's confidence and volunteer enthusiasm for the fight.

"We're off again, lads," said Darling, as he stepped out of the Dispersal building. "Another inbound raid." He glanced over at Koenig as the American came around the corner looking like death. Darling did not say anything, as they began to trot toward the fighters being started.

The late afternoon raid appeared to be a carbon copy of the morning attack. The Germans concentrated on the Chain Home RDF station at Ventnor on the Isle of Wight although they also looked for anything that might be an appropriate target . . . a fishing boat in the Solent, a Southern Railways train bound for the East, or the docks at Southampton. Brian thought of the Supermarine factory making Spitfires at Woolston across the River Itchen from Southampton. If the Germans hit that factory, they would soon be out of fighters. The thought gave Brian a greater, sharper edge to the ferociousness with which he engaged the Germans. Nonetheless, the results were the same. The aerial victories this day belonged to the Hurricane pilots among the bombers. A few fighters fell, but for the most part, the day was a draw for No.609 Squadron. Their sister squadron, No.152 Squadron, had not been so fortunate. They lost two pilots with three others wounded. Temporarily, they were near half strength in pilots with replacement machines in place or en route. It was a costly day in other ways.

Brian waited for the intelligence debriefings to be complete and the squadron to be released for the evening before he managed to find 'Slim' Koenig alone. "This is getting to you isn't it 'Slim'?" he asked.

"I'm OK."

"Which is why you puked your guts out before the last launch."

"I was hopin' no one saw that."

"We're all scared, 'Slim.' Some of us handle it differently."

"How would you know about scared? You're just a kid."

"Maybe, but I've been shot down three times and nearly died the last time, if it hadn't been for a woman who pulled my ass out of a lake."

Koenig remained quiet and pensive. Brian thought he wanted to talk, but did not know quite how to start. The younger American volunteer recognized the importance of the moment. If 'Slim' Koenig did not get a grip on his fear, he would not last long in the air.

"You've been through a lot already. You can make it," said Brian.

"Sure. How would you know? You didn't see what the Germans did in France. They swarmed over all of us. It didn't matter what we did. There were too many of them, and it seemed everything they did was right. They're too good for any of us."

"That can't be true. We've done pretty well against them so far."

"Oh yeah, but you haven't seen the worst of it yet. These guys are good."

"We can beat 'em."

"I'm runnin' outta places to run. They pushed us back across France like so many branches in a raging torrent. They've decided they're going to invade England, and there's nothing to stop 'em."

"We'll stop them."

Koenig laughed an empty, nervous, sardonic chuckle, as if he knew the answers and no one else did. They all wondered what was next, but most of them chose not to worry about what was ahead. They simply did everything they could to deal with the moment.

"You'll see," 'Slim' said finally.

Brian considered his words, but knew he was correct. "You've got to shake this off, 'Slim,' or you're not going to make it."

"I'm not going to make it no matter what I do. There is nowhere else to go. We are at the end of the line."

"I'm not willing to admit that. We're going to beat 'em."

"Your youthful exuberance is admirable, 'Hunter,' but you'll see. Now, I need several stiff drinks, so if you'll excuse me, I think I'll go get tanked."

Brian let the taller, older man walk off toward the front gate and the Officer's Mess. He stood by the corner of the Dispersal building until the slender American disappeared behind a building. Brian thought about the grim view espoused by 'Slim' Koenig. He wondered if his fellow countryman might be correct. After all, he had fought as a volunteer pilot in *le Armée de l'Air* from the very beginning of the Battle of France until the very end. Maybe he knew more than Brian did. The younger American refused to accept the defeatist view.

"He is not in good shape, is he, sir?" asked Corporal Jennifer Warren, startling Brian somewhat since he thought he was alone. "Didn't mean to give you a start."

"No problem, Jennifer. I just thought you had already left."

"He is not going to make it, is he?"

"If he doesn't change his attitude, I think you're right."

"Pity, actually.  All you American gentlemen have been extra special nice.  Pity to lose any of you."

"Like 'Hank' Maxwell?"

"Yes."

They both stood side by side staring off to some distant place alone with their thoughts.  Brian knew Corporal Warren was precisely correct.  He remembered the many times Malcolm Bainbridge had told him about attitude and its relationship to success.  If you believed you could do something, you had won 99% of the battle before it started.  The key was attitude.  'Slim' Koenig had lost his edge.

"I'd better get along, or I'll miss evening meal."

"Have a good night, sir.  See you in the morning."

"Good evening, Jennifer."

Brian walked a short distance with her before she split off toward the enlisted mess.  It was going to be many long days for 'Slim' Koenig.  Brian racked his brain to find a way to help 'Slim.'  He considered mentioning the problem to Squadron Leader Darling, but if he did, Koenig might resent the intrusion.  In the end, he decided to talk to Jonathan Kensington.  His best friend always had a pretty good view of the world.  The time had to be right.  They needed every pilot they could get, and they could not afford to lose anyone for any reason.

———

*Monday, 12 August 1940*
*Headquarters, Fighter Command*
*Bentley Priory*
*Stanmore, Middlesex, England*
*21:30 hours*

The day had started badly and got worse with each hour.  While Fighter Command did have its successes, the overall result and the view from the top were not good.  Darkness brought a respite in the brutal onslaught, but for the senior staff, the activity transformed from the direction of the air defense system to retrospective assessment and planning for the next day's action.

Group Captain Spencer organized the late night meeting.  Air Chief Marshal Sir Hugh Dowding led the meeting with Hogan from Intelligence, Leonard from Operations and Air Officer Commanding-in-Chief, No.11 Group, Air Vice-Marshal Keith Rodney Park, MC+Bar, DFC, joining them.  Park's group had borne the brunt of the day's attacks, and his headquarters was the closest.  Sir Hugh asked him to make the journey, so they could plan first hand.

"James, if you would be so kind to give us a summary of the day's events," Dowding said to Air Commodore Hogan.

"Certainly, sir." Hogan stood, picked up a slender wooden pointer and moved to the large wall map of Southern England. "We are still compiling the information from the day's operations, but from our first look, this was not a good day. The Germans attacked virtually every aerodrome in the South of England as well as two convoys forming in Portsmouth harbor. Convoys SNAIL and CABLE were hit hard with seven ships sunk at anchor and a score damaged. Of greater consequence, I suspect, they managed to strike every Chain Home site from Hawks Tor in the West to Bawdsey in the East. Worst hit was Ventnor. Reports, which I must say are incomplete, indicate Ventnor may be out of action for a fortnight or more. Several others including Pevensey and Rye as well as the Chain Home Low stations at Dover and Beachy Head will most likely be off for a day or so. The worst of it though is our loss of pilots. By our tally, this was the worst day in the history of the RAF. Twenty-three pilots killed or missing. Another score wounded with three very serious. Our counts indicate we lost 25 aeroplanes while the enemy lost 33."

"They can afford to lose that many," mumbled Leonard.

Hogan as well as the others ignored the quiet comment. "This was a well planned and executed attack over a broad range. Their reconnaissance and intelligence collection has been impeccable."

"Did they hit Woolston?" asked John.

"No. Miraculously, they did not. They bombed the docks at Southampton, but did not come close to the Supermarine factory."

"Thank God for small favors."

"Maybe the bastards haven't figured it out, yet," added Leonard.

"Whatever the reasons," Dowding said, "we were fortunate today. The day's attacks must give us pause to reassess our defense strategy. Until Lord Beaverbrook can bring Castle Bromwich into full production as an additional source of Spitfires, we must find a way to protect the Woolston plant. It is our only production facility for Spitfire fighters. The consequences of the loss of that factory should be painfully obvious to all of us. Keith, Woolston is in your area of operations. Is there anything you need?"

"In light of the attacks across my entire group, I shall certainly need support from Brand on the right flank and especially over Southampton as well as from Leigh-Mallory on the left flank."

Air Vice-Marshal Trafford Leigh-Mallory, CB, DSO, aerved as Air Officer Comanding-in-Chief, No.12 Group, covering the British Midlands.

"Geoffrey, if you would communicate with both One Oh and One Two Groups to ensure they appreciate the need for their fullest support of Air Vice-Marshal Park's Southeast group, it would be most appropriate."

"Yes, sir."

Dowding looked back to Hogan and nodded for him to continue.

"We also have seen new tactics from the enemy. As best we can establish so far, we think this is a special operations group, designated *Erprobungsgruppe*, or Experimental Flight Group Two One Zero. They fly uniquely configured Messerschmitt One One Oh fighters designed for high-speed, low-level, bombing attacks. This group approached at low altitude underneath a medium altitude conventional bombing raid. Within sight of the coast, they split into three groups to attack the Chain Home stations. They were well-disciplined, exceptionally well-trained and practiced, and highly effective. Our assessment is, we will most likely see much more of this group."

"They managed to get in this time without opposition," added Leonard. "The RDF operators know what to look for now. The next time won't be so easy."

Dowding held up his right hand to signal an end to the *braggadocio*. "We have several key actions to decide, gentleman, so it may be a long night." He paused to receive an acknowledging nod from everyone. "First, recovery of the Chain Home stations must be our highest priority. I know each of you fully appreciates the importance the RDF early warning network plays in our defensive strategy. We must execute our deception plan to mimic the signals from the damaged stations. The enemy must be led to believe they did not seriously affect those stations, and then we must apply every resource available to this country to get those stations operational as soon as possible."

"We might have some conflicts there, Sir Hugh," interjected Air Marshal Leonard.

"You let me worry about that, Geoffrey. When the first reports came in, I immediately called the Minister and the Chief of Staff. Since it is nearly 22:00 hours and I have not received a call from the Ministry, I think it is safe to assume they have briefed the War Cabinet and been given the authority we need."

"Yes, sir."

"To continue . . . Second, we must redeploy our resources south. All the evidence points to the Channel. We can expect an unrelenting and unprecedented level of aerial bombardment to prepare us for the invasion. I want to see that Keith's group gets the support they need."

"I should remind everyone, tomorrow has been designated by the Germans as, Eagle Day. We don't think it means invasion necessarily, but we are fairly certain it is a significant operational day for the Air Force."

"Eagle Day, ay," said Park.

"Do you need any other assistance . . . engineers or whatever?" asked Dowding.

"Virtually every aerodrome was damaged, but the ground crews appear to be coping reasonably well for the moment.  What I need is more fighter squadrons and air defense artillery, none of which are in plentiful supply."

"Understood, Keith.  We shall reassess our deployment strategy, now that we know the German intentions.  Geoffrey, if you would see to it, I shall be most appreciative.  James, if you would be so kind, please see to the RDF recovery effort.  We simply cannot let them know how much damage they have done or how important those stations are to us."

"Yes, sir."

"I should like to see the redeployment options as soon as possible in the morning, Geoffrey.  I shall also ask that you coordinate with the group commanders to ensure their support."

"As you wish, sir."

"We still have much work before us tonight, and we must let Air Vice-Marshal Park return to his command.  Are there any other topics?" John Spencer considered several possibilities, but in the end knew they would not be appropriate.  "Excellent.  Then, we are adjourned, gentlemen.  Have a safe journey, Keith."

Dowding stood and gathered up his papers.  They all stood in respect to the commander.

"If I may have a short word before I go," said Park.  Dowding nodded, and they retired to his office.

None of the remaining officers spoke as they left the conference room to their offices.  It would indeed be a long night.  For John Spencer, the pressure of staff work took a deeper toll than the rigors of life in the cockpit.  While the physical dangers were certainly greater, nighttime always brought a respite.  There were no boundaries to paper.  It just kept coming and the need for enabling actions seemed to grow larger by the moment.  He continually wondered what forgotten or overlooked action at Bentley Priory might irreparably affect the fighter pilots defending the skies above them.  It was an enormous and horrific burden to carry.

A knock of the door jam preceded Air Vice-Marshal Keith Park by an instant.  His expression was serious and tired, yet he managed a faint smile.  Group Captain Spencer stood as the senior officer entered.  "John, I need your help."

"Whatever I can do."

"Smashing. I know I am going to have trouble getting support from Leigh-Mallory, but I simply cannot whine and whinge about it. Do what you can to find the balance."

"No problem, sir."

"This is going to get quite nasty for many days, and the battleground will be the Southeast, not the Midlands. We are going to need all the help we can get to keep the friggin' Huns off our shores."

John always liked Keith Park's attitude and approach. Even as a New Zealander, he felt the same ownership and protectiveness they all did. He was cool, calm, collected, focused and committed as he had been in the cockpit of his Sopwith Camel during the Great War.

"You have my commitment, sir. I shall do my level best to help. I share your concerns, and only wish I could fill a cockpit for you."

Park laughed. "As do I. Brothers-in-arms, ay, John. I often think life was simpler then. Maybe a smidgen more bloody but simpler. Right and wrong were easy to determine."

"I still feel my skills, despite my age, are better utilized in the cockpit than behind a desk."

He chuckled this time. "Now, I share your views. I still feel more comfortable in the seat of a Hurricane than I think I ever will directing the air war in my area."

"Wherever I can be of service, sir."

"Thank you, John. I've asked for it. This may sound rather odd, but timely support is my greatest worry today."

"Yes, sir."

"I should be off. 'Til the morrow, then, John." Park turned to go and stopped at the door. "Please do tell your uncle, if you ever get to see him . . . he is doing an absolutely smashing job. Hell of a man, John, hell of a man."

"It has been more than a half year since I have seen him last, but I shall tell him upon the next opportunity."

Air Vice-Marshal Park left the Headquarters building to return to the demands of No.11 Group at Uxbridge. John Spencer knew Park as an exceptional leader of warriors and could not imagine anyone more qualified to lead Fighter Command at the phalanx. His reputation as a hands-on pilot who still flew his Hurricane everywhere and whenever he could find an excuse stood him in good stead with the line pilots in the seats today. John Spencer only wished he could be more help to the man, to his mission and to his group.

———

*Tuesday, 13.August.1940*
*RAF Middle Wallop*
*Middle Wallop, Hampshire, England*

**N**one of the pilots grumbled a word about Squadron Leader Darling's decision to withhold the normal 24-hour break straddled across the night every third day.  The air war plus the invasion alerts gave them sufficient reasons to accept the decision.  For Pilot Officer Brian Drummond, noon this day would have been his expected break.  Pilot Officer Roland Stockard gave his up yesterday and showed no detectable regret or resentment.  Brian could not avoid thinking about how long the ban might last.  He wanted to make a connection with Charlotte Palmer.  He needed the break.  Brian had been flying or on duty every day since being released from the hospital a week ago.  The bad weather days at the end of last week gave them a rest from combat but not duty.

No.609 Squadron came to Available status at 07:00 hours and was brought to Readiness at 09:15 hours.  No.11 Group to the east was already engaged in combat over several of the coastal aerodromes.  They waited for their turn.  The two Czech pilots wanted to join the action in the east, but they all sensed there would soon be enough for everyone.

The telephone rang.  'Slim' Koenig managed to keep his breakfast down.  "Six Oh Nine Squadron, scramble, scramble, scramble," came the shouted command from Corporal Warren.

The pilots were off and running as their engines began to fire off.  All the propellers were turning over at fast idle as the pilots reached their cockpits.  The last of the twelve Spitfires raised its wheels less than two minutes from the launch command.  As they climbed toward the Southwest having received their positioning instructions from Sector Control, they learned their target was a small raid heading for Portland, and they would have both fighters and bombers.  Darling decided his tactics.

"Sorbo Yellow, you take the heavies.  The remainder with us on the shooters.  Let's keep the bandits off the shipping."

No answer was needed as they found their target – a flight of six Ju88 medium bombers escorted by a squadron of Bf109 fighters.  They intercepted the enemy raid several miles off shore.  The fight quickly became what all the recent engagements degenerated into, a ball of twisting and turning fighters with only brief shots.  Blue, Green and Red Sections were nearly successful in occupying the escort fighters, although a couple managed to disrupt the efforts of Yellow Section.  The proximity of the fighters and ferocity of the engagement produced the desired result as the bomber crews lost their nerve, dropping their bomb loads into the sea short of their target, and then scattered and dove to gain speed and separation from the Spitfires.

Back at RAF Middle Wallop, numerous pilots reported hits, but none of them claimed any victories. The warm summer day mixed with the physical and mental exertions of the morning's aerial combat and sapped the usual animation. They drank water or tea, and then collapsed into any available chair with some on the grass under a small tree not far from the Dispersal building.

Brian heard the click of the loudspeaker and had just an instant of a thought about listening to a speech by the Prime Minister when the words came. "Six Oh Nine and Two Three Eight Squadrons, scramble, scramble, scramble." They leapt to their feet in a dead run. It was quite odd for a launch command to come over the loudspeaker system, but the effect was the same regardless of the method.

The 'VK' Hurricanes of No.238 Squadron made it airborne just ahead of the Spitfires of No.609 Squadron. Another raid on Portland. The unfortunate convoy trying to form up was not being given much of a chance, as the Germans kept up a seemingly constant assault before they could weigh anchor and get underway. This time the bombers were a squadron of Ju87 Stuka dive-bomber escorted by a squadron of Bf109 fighters. Brian told himself the Hurricanes would have a good time with the Stukas. The thought did not last long, as they prepared for hard work dealing with the fighters. Everybody saw everybody as they headed toward each other head-on at closure speeds approaching 700 miles an hour. A line of 12 Spitfires in a face off with a line of 16 Bf109s. As Brian pressed his firing button down, the flashes from their opponent's leading edge guns marked the beginning of the engagement. The first pass burst into the usual furball of darting, dodging fighters, as they each tried to find the right moment for a decisive shot.

For more than 30 minutes, the battle raged over the sea. 'Waggle' Davies was hit. His engine was smoking but still turning as he disengaged and headed toward RAF Warmwell. One of the Germans must have been hit and did the same thing. Then, miraculously, the German fighters disengaged *en masse*, but apparently did not communicate with the bombers. The remaining Spitfires dove to join the hunt on the grossly out-classed Stukas. Of the 16 dive-bombers that constituted the raid, only three managed to escape the slaughter.

Although Brian did not find a clear target during the entire engagement, he felt a peculiar nauseating sensation in his gut as they returned to RAF Middle Wallop. The strange feeling came from the enormous disparity between the British fighters and the lumbering dive-bombers. Fighter to fighter, the fight seemed fair, but the Stukas were like watching sheep being torn apart by a pack of wolves.

The first thing any of them wanted to do as they walked up to the Dispersal building was hear about Flight Lieutenant John 'Waggle' Davies. The report Corporal Warren gave told them he was on the ground safely. His engine seized on short final, and he landed a little hard without serious damage to the aircraft. They thought they would have him flying within a day of receiving a new engine. A Lockheed Hudson would pick him up later in the day and return him to Middle Wallop. Davies' good fortune did not alter the knot in Brian's stomach.

His mood from the end of the mid-day mission carried through the debriefings and his feigned slumber as they waited for the next launch. He did not feel like talking about it, but the sensation persisted. He also found it quite odd as he wished the Germans would not use the Stukas. If there was going to be a fight, he wanted it to be a fair fight like the free-range pheasants he and Malcolm hunted nearly two years ago.

The fatigue soon took him into an odd slumber as his body and mind tried to recover some strength expended in the two fierce combat sorties. A light breeze came up to offer just enough cooling to make the Sun's heat relax the muscles. The knot slowly began to unravel. The quiet around him other than the gentle rustle of the leaves allowed him to oh so briefly separate himself from the intensity of operations at the important Hampshire aerodrome and sector station. Maybe the war would simply go away and give them some rest.

Brian sat up as if he had been jolted with a strong electric charge and each of his brethren reacted the same to the strange wail of the siren. He had heard the sound in London, but never at Middle Wallop. They all searched the sky wondering if it was a malfunction. Just as he and others caught the dark, angled, crank-wing shape of the Stuka dive-bombers, someone screamed, "Scramble!"

Chairs flew through the air as the pilots hurtled toward their fighters. Ground crews scrambled from the shade of wings in a frantic effort to get engines started. As the sparks ignited the fuel-air mixture in the large cylinders of several Merlins, the eerie and morbidly distinctive shrill scream of a Ju87 *Sturzkampfflugzeug* filled the air.

"Don't wait," someone shouted over the cacophony around them. "Go, go, go!"

Pilot Officer Brian Drummond did not wait. Bombs began exploding all over the airfield as Brian pushed his throttle forward as fast as he could and still hold his nose reasonably straight. Brian's Spitfire rumbled across the grass clawing for air. He felt the wheels go light and eased back on the stick as the machine took to the air. He reached for the landing gear handle as a large bomb exploded directly in front of him. The fighter rocked up on its left wing as the metal rang with impacts from shrapnel, dirt and other debris. The engine coughed several times as it gasped for clean air. Brian had the stick full over against the right stop, and it continued to roll left. He knew he would soon be on his back with the hard ground about to grab him. As the right wing pointed to the sky, the wounded Spitfire began to respond to Brian's commands. She shuddered and shook trying to respond short of stalling, and she did. He was in clean air and the Merlin roared back to angry, defiant, full power.

Brian allowed the bird to accelerate holding her down just above the treetops as he quickly flipped his switches up arming the eight, fully-loaded, Browning machine guns. He passed 200 miles an hour, pulled back on the stick and rolled the 'PR-F' Spitfire slightly to find a target. The smoke and plumes from exploding bombs occupied the space behind him that was RAF Middle Wallop. He saw the Stukas still diving on the airfield. The distinctive black, oil smoke with its embedded orange flame told him at least one or more aircraft had been hit. He continued his turn toward the nearest dive-bomber, and then instantly diverted his attention elsewhere. Where were the fighters?

Brian scanned the sky above many times in the span of a few seconds. There were no fighters in sight. They could be behind the few clouds, but they had to be somewhere. He could not wait for them, so he decided to make them come to him. He found a Stuka with its nose slightly high and rolling to enter its dive toward the smoking and damaged airfield below. He rolled gently to the left and moved his sight pipper slightly ahead of the Stuka's propeller dome. He closed rapidly and held his sight until it seemed he might hit the unsuspecting attacker. He squeezed the red firing button on his circular spade on top of his control stick. The eight guns responded instantly spitting a concentrated stream of brass-encased lead bullets into the German airplane. The dive-bomber evaporated into a cloud of tiny parts and chunks of wing, tail and a mass that was a Daimler-Benz engine. Some of the shrapnel peppered his Spitfire, but she was still flying.

There was no time to savor the moment or worry about what might be. He searched the sky again for fighters. None, yet. Brian rolled back right nearly inverted, as he pulled his nose through and looked for another target. He quickly found another one just as oblivious as the previous one. The enemy pilots must be very intently focused on their target. He adjusted his flight path. He had an excellent angle for what would soon become a head-on shot. Brian checked his slip ball . . . good alignment. He was closing fast, but still not in range, and then the German saw him.  He rolled his slow aircraft away from his target line and Brian's rapidly closing angry Spitfire.  It was too late.  A small adjustment.  Brian opened fire sawing the Stuka's tail off.  The mortally injured German airplane began a grotesque spinning descent to the middle of the airfield below.  Brian did not wait for the impact.  He looked for another target.

The other Germans, undoubtedly alerted that at least one British Spitfire was airborne and not very happy, wisely decided they had done enough damage.  He found several diving for the treetops heading south.  Two Hurricanes and another Spitfire were now in the chase and were closer to the retreating bombers.  Brian searched the entire sky for fighters.  There were only small cumulous clouds.  He appeared to be the only fighter with height and his colleagues seemed to have a good bead on the Germans.  Brian decided to stay up, just in case the fighters might appear out of nowhere.  It would be up to him to defend his brothers below.  He searched and searched, but there were no fighters.

Brian allowed himself a few quick glances below to watch the Stukas be torn apart by the clean, purposeful strokes of the British fighters.  He continued to look for the fighters and could not avoid the questions revolving around why, after the Stukas had taken such a beating over the last few weeks, the squadron of Ju87 dive-bombers showed up at a British fighter base without fighter escort to protect them.  Where were the warning calls?

Just then, Brian realized he did not have his helmet, headset or oxygen mask on, nor could he find them.  He thought he might have dropped them in the rush to get off the ground, but they were nowhere to be found.  Even more surprising, Brian was not strapped into the aircraft.  He did not have his shoulder harness or lap belt fastened.  He quickly gathered up and locked at least his lap belt.

The other fighters eventually came up to join him as he circled the airfield.  They pointed toward their earphones signaling him to put his headset on and listen to the radio calls.  He held up both hands and shrugged as if to say there was nothing he could do.  The engine's coolant temperature began rising despite the low power setting, a wide-open radiator and cool temperature at altitude.  The bird had to be losing ethylene glycol coolant.  Brian signaled the others he had to land.  They shook their heads to tell him not to, but Brian ran a hand across his throat and energetically motioned toward the ground.  He did not have much time before the coolant ran out and the engine froze.  At that point, he would have no choice.

Brian rolled his 'PR-F' Spitfire and pulled the nose toward the ground. He made a tighter orbit as he evaluated the landing area pocked with craters and several burning aircraft. Airplanes that could not get off the ground were scattered all over the airfield. He found a reasonably acceptable line close to the wind line. As he circled to land, he saw what was behind the smoke. Hangar No.5, his squadron's hangar, had a large hole in the roof and was burning with several of the large, steel, sliding doors blown completely out, laying on the ground in front of the hangar. The damage at the airfield appeared to be substantial and serious.

He landed and tried to taxi back to their line. He made it as far as the Operations building when the obstacles made movement more difficult and the coolant temperature reached the redline. He shut the engine down.

Leading Aircraftman Bernie Gordon ran through the smoke and debris toward him followed by Toldson and Jenkins. Brian was out of the cockpit, down the wing and on the ground when Gordon made it past the left wingtip. "We saw everything, sir. It was fantastic," Gordon said.

"Best show I have ever seen," shouted Toldson.

"We saw you go and nearly lose it in the explosion just beyond the crown of the landing area. We could not see you again until you began your climb, and then we saw you go after the bloody, frigging Huns."

"First, one."

"Exploded," interjected Jenkins.

"Then, the next one."

"The bastards crashed on the southeast corner of the aerodrome."

"What happened in the hangar?" asked Brian.

The mood changed instantly from effervescent excitement to somber recollection. "One of Gerry's bombs hit square in the hangar and blew the doors off. The chief thinks there may be two of our blokes under them. The fires are out, but the damage is pretty extensive."

"How many got airborne?" Brian asked, wanting to change the subject.

"You, along with Misters Beamish, Morrow, Kensington, Burns and Kradilcek. Squadron Leader Darling was wounded in the arm and leg before he reached his aeroplane. Mansek and Davies were also wounded. Koenig had his aircraft destroyed before he got there. All in all, not a good day for us I'm afraid. The chief is still counting our casualties, but before we saw you circling to land, the count was at least three killed, maybe more, scores wounded, and we have lost at least three aeroplanes. How is . . ." Gordon stopped, as he looked at the Spitfire fighter. He crouched down to look underneath.

"Dear God," he added, as he crawled toward the aircraft, and then flipped onto his back to scoot under the airplane beneath the cockpit. "I have never seen so much damage to an aircraft." As Brian and the others joined him, Bernie Gordon continued his assessment. "Coolant and lubricating systems look totally gone. There are holes across the entire underbelly. Maybe some serious structural damage."

"How long?"

"Hard to say, Mister Drummond. She may never fly. The engineers will need to look at this damage before we can tell if she is even repairable."

"I'm not sure we should leave her here."

Station Operations Officer Wing Commander Herbert Worly walked over to Pilot Officer Brian Drummond. He extended his right hand to Brian. "Smashing work, Mister Drummond," the senior officer said the first words to the American volunteer. "Brilliant piece of airmanship, simply brilliant. No wonder you are an ace."

"Thank you, sir."

Gordon came out from under the injured Spitfire. He saluted the wing commander, as the other two crewmembers did as well. "Took off in such a rush, he forgot his flight kit, sir."

"Probably forgot to buckle in," added Toldson.

Brian offered a feeble grin. "Yes, I didn't have my flight kit nor was I buckled in, and I never realized it until the fight was all over," Brian muttered.

"Well, my boy, probably the difference that got you in the air when others did not," said Worly. "Nonetheless, fanciest bit of flying I've seen. You should get a medal for what you were able to do," he said shaking Brian's hand again. "Oh, my, young man, I might add, the tail of the Stuka you managed to cut in half alighted near the south boundary. We shall see if we can carve out an appropriate souvenir for you."

"Thank you, sir."

Wing Commander Worly nodded his head and returned to his building and his tower. The heavy, stinging odor of burning fuel carried over the airfield from the hulks across the landing area. The fire brigade methodically extinguished the fires as crews began the arduous process of clean-up and repair.

"You'll be famous after that show," said Bernie Gordon. "We shall start working on your fighter," he said, using the first reference to Brian's possession of the Spitfire rather than his own. "It will probably be a few hours before we will know how badly she is damaged. I will get the engineers on her as soon as I can find them."

"Great, Bernie.  Thanks, guys.  I'll be over at Dispersal when you get a measure."

Brian walked around several craters, incalculable debris spread about by the explosions as well as several still smoking, partially burnt remains of Hurricanes and Spitfires.  The damage done by the German surprise raid was enormous.  He considered the possibilities of how the squadron of German dive-bombers managed to make the trip to Middle Wallop without being spotted or engaged especially without fighter cover.  They had to know more about the British defenses to find any seams or cracks in the system, and they had to come in very low.  The raiders clearly knew exactly where they wanted to be, but paid a terrible price to partially accomplish their mission.

The airfield appeared to be swarming with people, Air Force and civilian, pulling wreckage off the landing area even before it had cooled off.  They busied themselves filling in the craters, packing and smoothing the surface, and picking up any debris that might cause damage to fighters taking off or landing.  The concentration on the grass field meant the sector station wanted to return to full operations as quickly as possible.

More than two-dozen fighters now circled the airfield in several layers undoubtedly with instructions to stay aloft as long as they could to protect the field from any further attack as well as allow the crews to accomplish as much repair as possible before they landed.  Brian found several of the pilots, Stockard, Johnson and Koenig, sitting outside the Dispersal building watching the bevy of activity around them.  Corporal Warren stood in the doorway of the empty building.

"Good job, ace," said Koenig, as Brian approached.

"Smashing, absolutely smashing, Brian," added Stockard.

"Thanks."

"We all thought you were done in with that explosion during your takeoff," Jennifer Warren said.

"We missed it," said Johnson, "but, everyone was talking about it.  Looked like you single handedly chased off the bad guys."

"I just reacted the best I could and was real lucky."

"Better lucky than good, as we say," Stockard said.

"Squadron Leader Darling, Flight Lieutenant Davies and Mister Mansek were wounded and are being treated at Surgery," offered Warren.

"Bernie Gordon told me," answered Brian.  "Has anyone been to see 'em?"

"Not yet," responded Pilot Officer Roland Stockard.  "We haven't been released yet, even though we have no aeroplanes.  Corporal Warren has checked on them.  Everything seems to be stable."

The others began to return, parking their machines near the Operations building. Brian's Spitfire had already been towed off to one of the other undamaged hangars to start the evaluation and repairs. As the pilots gathered, they talked as they always did about their flying experience that day whether personal or through others. This day they had an enormous amount to talk about. The weather in the east had been worse all day, but that did not stop the bombers from coming. No.11 Group experienced the same inexplicable visitation by unescorted bomber formations that were decimated by the British fighters although sufficient numbers penetrated to their targets. Damage was substantial. Manston, Biggin Hill, Kenley and even Northolt had been hit hard. To the pilots of RAF Middle Wallop, the day seemed most severe to them. The war had become much more personal and real.

Brian remembered the stories Malcolm Bainbridge told him about the necessary ruthlessness of war. Kill or be killed, he had said. Brian's understanding of the phrase was clearly defined on this Tuesday in August. He made the transformation from pity and remorse, witnessing the slaughter of the considerably less capable German dive-bombers to the surgical elimination of those same aircraft threatening his home.

By the time the pilots retired to the Officer's Mess, the consequences of the enemy raid had been determined. Seven dead including the three men under the Hangar No.5 doors, 33 wounded including eight pilots, and five aircraft destroyed and a score damaged to various levels. It had been a very costly day for RAF Middle Wallop, but there was only one consistent topic – Pilot Officer Brian 'Hunter' Drummond's incredible exhibition over the airfield. True to his word, Wing Commander Worly entered the bar well into the evening to present a souvenir to Brian with considerable fanfare. The tail swastika from Brian's second Stuka dive-bomber that day had been cut out of the wreckage. It was now a war trophy in the same tradition as *Rittmeister* Manfred *Freiherr* von Richthofen – Captain of Horse Manfred Baron von Richthofen – the bloody Red Baron.

---

# Chapter 4

Man's inhumanity to man
makes countless thousands mourn!

-- Robert Burns

*Wednesday, 14.August.1940*
*Poole Harbor*
*Poole, Dorset, England*
*11:10 hours*

### Week 6

The Short Brothers model S.30, Empire flying boat was painted in the livery of British Overseas Airways Corporation and was named Clare. It gained speed to the west across the harbor, north of Brownsea Island, into the day's westerly wind of 10 knots.  Once safely into the air and climbing, the flight crew set course for Montreal, Canada.  The flight was planned for and scheduled to take 25 hours to complete, give or take, depending upon the actual winds and weather encountered as well as the turnaround time for fuel and provisions at their planned enroute stops at Foynes, Ireland, and Botwood, Newfoundland.

Among the 19 passengers and seven crewmembers, the aircraft carried two British citizens setting out on a very important mission for His Majesty's Government -- Sir Henry Tizard and Group Captain Franklin Pearce.  They were the lead element -- the vanguard -- of the British Technical and Scientific Mission to the United States.  They would have three weeks to finalize the agenda, attendance, transfer process and logistics details with their American counterparts before the main body of the British delegation arrived in Washington, DC, by ship and rail.  High hopes accompanied them.

---

*Wednesday, 14.August.1940*
*Headquarters, Fighter Command*
*Bentley Priory*
*Stanmore, Middlesex, England*
*13:50 hours*

The telephone rang in the office of Group Captain John Spencer, coming as relief rather than annoyance.

"John, you might want to join us," said Air Marshal Leonard.

"Yes sir."

"Gerry appears to be taking another big day, actually.  The board is brimming with inbounds."

"I'll be right there," John answered.  The rationalization for his visit to the underground Operations bunker as a request from a senior officer rather than an excuse to avoid his paperwork.  John knew the paperwork was important, but he would never enjoy the process.

The action over the last fortnight made the agony of his confinement to the office all the more painful.  Although he was certainly not as young as he needed to be, John wanted to feel the rush, challenge and satisfaction he remembered from the glory days of the Royal Flying Corps in the Great War.  He neatly restacked his undone work, told his assistant where he was going and headed toward the entrance stairway descending underground.

John joined several other members of Fighter Command's senior staff, including Air Chief Marshal Sir Hugh Dowding in the gallery above the massive map board of Great Britain and the distant buzz of many muffled conversations.  At first, no one acknowledged his arrival as they intently watched the enormous number of enemy raid marker blocks as well as growing number of friendly blocks moving to meet them.  The plotters gracefully pushed the blocks across the table with each update from the Filter Room.  John had never seen the Operations Room so busy even during the rehearsals and practice during the Phony War.

Air Marshal Leonard leaned toward John and whispered, "They appear to be quite intent upon disabling our fighter aerodromes, I'm afraid.  Half the facilities in One One Group are already under attack, and as you can see, the other half will most probably be engaged shortly."

"When did it start?"

"Mid-morning, unfortunately.  This is the second wave and the largest we have seen to date."

"How about the other Groups?"

"One Oh Group is not quite there but close.  Gerry seems to be focusing his attention in the West Country upon Middle Wallop.  That's one of our primary sector control stations especially in the West.  Makes one wonder how much they know about our air defense system, doesn't it."

"Damn!"

"Sorry?"

"So sorry, sir."

"Oh, yes," Leonard said with his recollection.  "Your young American bloke.  They must recognize the location of one of our best squadrons.  Six Oh Nine is our third highest scoring squadron at this point as well as having the most American volunteers along with two Czech pilots."

"Yes, sir."

Leonard rubbed his chin, as if that would help his memory.  "Drummond, isn't it, John?"

"Yes, sir."

"I noticed in yesterday's results he was credited with his eleventh and twelfth victories."

"Yes, sir."

"From the reports, he nearly single-handedly beat back the bombers attacking his aerodrome.  Rather impressive display of airmanship overhead the facility, so they say."

"So I hear as well, sir."

"Has he received his Distinguished Flying Cross, yet?"

"Not that I am aware of."

"The lad deserves the Military Cross, if the report withstands a proper review," offered Air Marshal Leonard.  "Brilliant, most impressive, I'd say."

"If we can just keep him alive," John mumbled to himself.

"Pardon?"

"Nothing, sir.  Just thinking aloud, I'm afraid."

Their attention turned to the map board.  They watched and listened intently to the communications between the Fighter Command Operations Room and all four, fighter groups.  While No.11 Group was the most heavily involved with No.10 Group not far behind, German raids were over targets in every area of operations except Ireland.  The battle continued for thirty minutes until the first signs of withdrawal began to be indicated on the map board.  Impromptu damage reports provided a measure of the effort made by the Germans.  Several airfields – Manston, Dover, Biggin Hill and Tangmere – were temporarily out of action due to cratering of the landing area.  The fighters were being diverted and recovered at secondary airfields.

As the enemy raids subsided for the moment and no new raids appeared to be inbound, the Air Officer Commanding-in-Chief left the Operations Room without a word to anyone.  John could not help but wonder what his commander was thinking.  Every day was a trial of life, of survival, of freedom.  The reports did not sound good.  The losses were chilling, especially the losses of pilots that were much harder to replace than the aircraft.  Although just as the fighter aircraft production had taken a positive turn, disappointment, as was so often the case, rode along with the few successes they enjoyed.

"Gerry finally figured out where our fighters are built," began Air Marshal Leonard, as if the discovery had just come to him.

"They hit Woolston?" asked John with surprise and fear.

"No, not yet, thank the dear Lord above.  They bombed Castle Bromwich last night, just as they were beginning to deliver their first Spitfire Mark II.  Serious damage but nothing critical."

"It is beyond my comprehension," said John, "why they have not struck Woolston, virtually our sole source of Spitfires."

"A puzzlement to us all, John.  A most beautiful twist of fate, I should think.  They hit the Hurricane factory at Brooklands.  They have also hit Eastleigh, the Supermarine flight test facility, which may be why they have not quite figured out where the bloomin' things are built."

"Thank God for small favors."

"Indeed."

"While I am here, I shall use any excuse to avoid that most abominable stack of paperwork in my office.  What other excitement have we had?"

Leonard chuckled with empathy for the Staff Secretary and considered the question.  "Are you aware of Bomber Command's extraordinary effort last night?"

"No."

"Small group of Wellingtons, a half dozen I think, flew to Italy last night to offer a most appropriate greeting to *Il Duce*.  They dropped their bomb loads, albeit reduced loads for range, on the Caproni and Fiat aircraft factories in Turin and Milan.  Good hits according to the Air Ministry."

"That must be more than 1,500 miles round trip – near the max range for a Wellington – maybe seven or eight hours in the seat."

"I do believe."

"At least, we are giving them a taste of their medicine.  I wonder how they like it?"

Both men laughed trying to find some brightness in their rather dark situation.  Before the last wave of raids disappeared from the table, new formations began to appear at the southern reaches of the Radio Direction Finding equipment.  The new raids remained stationary as they circled above their airfields gaining altitude and allowing other aircraft in the formation to join.

"How many Chain Home stations are still out?" asked John.

"We managed to reinstate most of them during the night.  We are still missing Ventnor and Poling.  The latter should be operational later today or tonight."

"That leaves Southampton without much warning."

"Afraid you're correct, old boy.  The chief has asked us to keep a squadron on patrol over the Isle of Wight until at least Poling comes back on line and maybe Ventnor as well, depending upon how Gerry plays it."

John Spencer listened to the gradually expanding murmured voices of the plotters, as he peered down upon the air defense map and pictured the gap in their early warning system.  If the Germans only knew how fragile, vulnerable and vital the Chain Home Radio Direction Finding stations were to the Air Defense System of Great Britain, they could render Fighter Command blind.  The loss of warning and preparation time would cause them to expend far greater resources in terms of fuel and precious pilot assets to compensate.  The process of grinding down the fighter defenses would proceed at a greater pace than it already was.  Everyone with knowledge of the situation from the Prime Minister to the pilot sitting in the lonely cockpit had to be asking themselves the same question – how long could they hang on?  The protection of inclement autumn weather seemed so far away.  When would the Germans find their Achilles Heal, or more appropriately, when would they recognize that they already found it?

"Winter seems so far away," mused Group Captain Spencer.

"We are all thinking the same thing, John.  We simply must determine the successful way forward, and with a little divine providence, we shall find it."

They watched as the building raids began their track toward Great Britain.  John considered all the information about German tactics and intentions he could remember.

"Didn't the last intelligence report tell us their so called Eagle Day was the 13th?" asked John.

"As I recall."

"Maybe we've seen the worst of it."

"The spooks don't think so.  They say the size and frequency of the raids against their known order of battle and resource estimates indicate we have only seen a fraction of what could be their maximum effort."

"Isn't that a pleasant thought?"

Air Marshal Leonard only nodded his head, as his attention was drawn more magnetically toward the many H – for hostile – data blocks moving toward Great Britain.  The large multi-colored clock on the far wall said it was late afternoon.  The number of hostile blocks looked like about a quarter of the previous wave and probably meant the last major assault for the day.  Only the few nighttime raiders would be left to harass them until the next dawn.  Time was beginning to have meaningless value, as the sanctuary of foul weather seemed to move farther away.  After offering further guidance to several of his sector controllers to be passed along to the actual sector controllers, Leonard looked briefly at John, and then back to the board.  "Sir Hugh predicted this would be the decisive phase.  He also acknowledged the fact that we have insufficient resources. We cannot muster sufficient fighter assets to protect just One One Group's aerodromes, set aside the other groups and manufacturing facilities." Leonard glanced at John Spencer again and received only a head nod of recognition. "To be rather crassly blunt, we are losing pilots at an alarming rate, and at that rate we will not be able to make it to the winter weather."

"I have seen the loss reports as well. We need a miracle."

Leonard chuckled slightly in a nervous sort of manner. "To say the least, I should think."

"What do you think we should do?"

Leonard locked onto John's eyes, as if to ascertain his sincerity. "First, we must find a means to distract or divert the German attention on their plan. Until then, we have no choice but to keep One One and One Oh Groups at full strength, which means keeping those pilots in their cockpits as long as the threat remains viable. We must make up losses by drawing down One Three Group, and then One Two Group."

"Leigh-Mallory will not look kindly on that endeavor."

"Fuck him!" spat Leonard, in an unusually rare burst of profanity. "My apologies, John, that was not particularly gentlemanly of me."

John Spencer smiled and said, "Not to worry, sir . . . well understood."

"Have you heard the latest from that malcontent?"

"You mean Bader's Big Wing proposal?"

"Yes, precisely. The bastards act like they have or should be given sufficient time to form up several squadrons prior to engaging the enemy formations. From his distance in the Midlands, he is probably correct, but we cannot give Keith Park that time. Chain Home cannot see over the horizon, and the farther away, the higher the altitude required to see them. It just does not work the way he wants it to work."

"What do you think he will do, if we did start drawing his group down?"

"He will scream bloody murder to the Chief and the Minister, and after that the War Cabinet and perhaps even the Press."

"Surely he wouldn't."

"I should not speak ill of a brother officer, John, but he bloody well cares about only himself."

John could say nothing in response. He knew Leonard was on volatile ground. John wanted to change the subject. "There are bright signs."

"Bright?"

"Well, perhaps not bright, but encouraging."

"To what are you referring?"

"I saw a report this morning the testing lads flew a Beaufighter equipped with a Mark Four RDF unit. I think they are now calling it RADAR, which stands for RAdio Detection and Ranging. They successfully found, tracked and closed with all three of their presented targets. The results are very encouraging. They are still testing the unit, so they have not worked out the engagement tactics."

When did this happen?"

"Two days ago, by the report."

"I sure hope they are quick with their development work. We need several squadrons of those machines yesterday."

"Yes, quite right," John said. "Now, regrettably, I should return to my paperwork before they bury me in it."

Leonard chuckled softly, again, without taking his eyes off the board. "Good luck, John. Thank you for listening. I trust you recognize our conversation was in confidence."

"Yes, sir, absolutely."

"Excellent."

Group Captain Spencer left the Operations bunker and return to his office. He knew Air Marshal Leonard was correct in his assessment and perspective, but he could not see the path forward. They all knew something had to change . . . only what?

—

*Thursday, 15.August.1940*
*RAF Middle Wallop*
*Middle Wallop, Hampshire, England*
*06:00 hours*

The devastating surprise attack of Tuesday caused many reactions within the fighter defense system. The resident squadrons dealt with their vulnerabilities. For No.609 Squadron, their new dispersal point was several large tents in the southwest corner of the facility. No.152 Squadron was more or less permanently relocated to RAF Warmwell, which gave Brian and his fellow aviators some solace for leaving a permanent building for a tent – at least they were in a tent at Middle Wallop rather than Warmwell. No.238 Squadron joined them moving into tents in the southeast corner leaving only the night fighters of No.604 Squadron at the main area on the northern side of the landing area.

The squadron was returning to strength rapidly. Squadron Leader Darling's wound had been minor, and he was back to full duty. Pilot Officer Mansek would return to half-day duty by noon, and Flight Lieutenant Davies would remain in the hospital several more days.  Davies' leg wounds were rather serious requiring a multitude of stitches to close.  The doctors estimated it might take a fortnight for him to have acceptable function in his leg and allow him to return to flight status.  Numerous cuts and bruises marked the severity of Tuesday's combat for most of the pilots.  Darling decided to give Mansek a two-day leave to recuperate and gave Davies a week pass to go home for his recovery.  'Fog' Johnson would fly in Mansek's position as left wing in Darling's Blue Section.  'Slim' Koenig would remain on duty as a relief pilot.

Brian's 'PR-F' Spitfire was expected back from repairs sometime in the morning.  Two replacement aircraft for the three destroyed machines had been delivered, painted, tested and declared ready for operations.  They were still one aircraft short, but with Mansek and Davies off the duty list, they had a spare.

The squadron was brought to Available-in-30-Minutes status before they finished their early breakfast.  Most of the pilots were barely alert as they wolfed down eggs, bacon and toast.  They stopped briefly at Available around 07:00 hours and went to Readiness at 07:10 hours.  Initial raids on the Southeast Chain Home stations had already begun although certainly not with great intensity.  No.11 Group fighters were airborne on their first sorties for the day, while the No.10 Group pilots waited their turn.

The squadron received a message from the Air Ministry at 08:05 hours.  They passed it among the pilots.  The message produced the same result with each reading – a sarcastic grunt or chuckle.

---

### SECRET

```
DATE 15 AUGUST 1940
FROM HEADQUARTERS FIGHTER COMMAND
TO ALL GROUPS STATIONS AND UNITS
RESEND
DATE 15 AUGUST 1940
FROM AIR MINISTRY
TO ALL COMMANDS
SUBJECT ENEMY AIR OPERATIONS
BEGIN
EXPECT HEAVY ENEMY AERIAL BOMBARDMENT ACTIVITY
COMMENCING 15TH AUGUST ESPECIALLY IN SOUTHEAST
AND ALONG CHANNEL COAST BREAK MAXIMUM READINESS
SHOULD BE MAINTAINED
END
```

### SECRET

---

"As if we haven't seen heavy bombardment, yet," joked 'Red' Burns. "We've seen worse than anything I saw in France.  I wonder what heavy means."

"It means a bloody shit pile full," barked 'Boxer' Stockard.

"It's already begun in the Southeast," 'Slim' Koenig said.

"So, they say."

This day continued to be distinctly different as well as agonizing for the pilots of No.609 Squadron.  The reports from the East told them No.11 Group was probably the objective of the heavy bombardment.  Several large raids split into smaller groups each heading for nearly every Southeast fighter station.  The information placed an unexpected pressure on the waiting crews like a condemned man awaiting his executioner.

They waited in silence scanning the hazy but cloudless, quickly warming summer sky.  All the engine run-ups had been completed hours ago.  The only sounds other than the birds chirping were the occasional rivet gun or other metallic clanking and clattering of ongoing repairs.  The sky remained dreadfully empty as the reports from the East mounted toward a very ugly image.

Brian speculated to himself about the German intentions.  Maybe they would concentrate on No.11 Group in an attempt to defeat them first to support their invasion plans, and then turn their attention to No.10 and No.12 Groups on either side of Air Vice-Marshal Park's busy organization. Would they be moving east to fill in for decimated squadrons?  Would the airfields be rendered useless for any aircraft, forcing the fighters to fly greater distances to counter the invasion landings?  When would the messages come that the Germans had landed at Brighton, Worthing, Eastbourne or Hastings? Would they hear the first signs the enemy had forced through the defenses at Dover, or that tanks were rumbling across the English countryside toward London?  They waited for the unknown.  They waited in silence.

The telephone rang and every available pilot shot to instant alertness as the message came.  "Scramble the squadron," announced Corporal Jennifer Warren.

They ran to their living, breathing fighters poised for immediate takeoff.  Pilot Officer Brian Drummond started to the scattered line with the others only to pull up short upon remembering Bernie Gordon had not yet reported his aircraft ready for flight.  Brian watched eight Spitfires taxi swiftly to the Operations Tower at the top of the small hill, and then roar across the grass passing over the boundary fence just in front of him.  He belonged with them, but his mighty steed was lame.  Brian would have to wait some more.

*Thursday, 15.August.1940*
*No.10 Downing Street*
*Whitehall, London, England*
*16:30 hours*

As the reports poured in, they all knew this was Göring's *Adlertag* – Eagle Day. With each new update, Winston felt himself become more irritated and annoyed. The last place he wanted to be was in a now routine afternoon War Cabinet meeting. Sir Edward Bridges, Secretary to the War Cabinet, recognized all the signs of the Prime Minister's progressively agitated state. The Prime Minister wanted to be near the action. With the windows open to allow what breeze there was to offer some cooling on the summer day, they could occasionally hear the explosions of bombs at RAF Croydon or some other aerodrome. The rhythmic, pulsing drone of German bombers passing overhead toward the northern perimeter aerodromes added to the poignancy of the Prime Minister's frustration. Sir Edward did his best to find the appropriate conclusion to the afternoon meeting.

With the pronounced adjournment, Winston bolted from the room. He barked to his Principal Private Secretary, John Martin, to call up the automobile as quickly as possible. He also requested that General Ismay, Military Assistant to the War Cabinet, accompany him. Of course, these days, wherever Churchill went his bodyguard Inspector Walter Thompson, accompanied him usually in the front seat beside the driver.

"Where to, sir?" asked the driver.

"Uxbridge. One One Group Headquarters as fast as you can," Churchill said sharply.

They rode in silence as the horn blared and the lights flashed to gain passage through the crowded afternoon streets toward Air Vice-Marshal Park's operations bunker. The primary action, as all the reports indicated, lay squarely upon No.11 Group. Winston sensed the pivotal character of this mortal battle and was nearly beside himself with anticipation although not a word was spoken during the half-hour journey to the center of the battle ... well, other than Thompson's grumbled commands to avoid traffic or other obstacle between them and this destination. Thompson usually defied Churchill on the prime minister's penchant for speed, but this time he sensed a genuine need and did what he could to help the driver navigate the course.

The Prime Minister did not wait for any of the usual courtesy or ceremony, as he bolted from the automobile before it had fully stopped. The entrance guards fortunately recognized the man who ran passed them. They saluted but too late. Churchill knew the way into the underground bunker.

---

*Thursday, 15.August.1940*
*Headquarters, No.11 Group*
*Uxbridge, Middlesex, England*
*17:45 hours*

**A**ir Officer Commanding-in-Chief, No.11 Group, Air Vice-Marshal Keith Park stood as well as several other senior officers in the gallery above the Southeast England operations map board. Winston waved for them to carry on and ignore him, as if he were swatting an annoying insect. He sat down allowing the others to sit as well. They watched silently for several minutes. Winston absorbed the mass of blocks moving across the map and the humm of muffled, professional, human communications as the struggle for air supremacy over the probable invasion beaches boiled in the skies.

Park leaned toward the Prime Minister sitting next to him and spoke softly. "They have struck every aerodrome in my area. Manston has taken a terrible pounding and is out of action. The squadrons have been diverted to other aerodromes. They are coming faster than I can refuel and rearm the fighters. We have lost numerous fighters caught on the ground trying to refuel and get back up in the air. I have asked for assistance from both Ten and Twelve Groups. We have also lost the Poling Chain Home station again, which hopefully will be operational by the end of the day."

The Prime Minister nodded his head in acknowledgment. He listened intently to the radio chatter from several sector control stations. The desperate battle raged on through the late afternoon and into the still bright evening hours. The swirling mix of friendly and hostile blocks on the board boggled the mind and could barely reflect the confusion and desperate battle in the skies above Southeast England. The numbers on the blocks that he had become quite familiar with told him the rapidly dwindling numbers of friendly fighters were being overwhelmed by the seemingly endless stream of German bombers and fighters.

Although not shown on Park's operations board, Winston received message reports from the Air Ministry that every fighter group in the Royal Air Force was now involved in the Battle of Britain. Raids from Norway had arrived late in the afternoon, attacking targets in Scotland and Northern England.

Against enormous odds, Winston Churchill observed the greatest air battle in history and the courage of young men bound in the confines of the small, cold, lonely cockpits. He watched as many rose for their fourth long combat mission of the day. The enemy was unrelenting, persistent, precise in their targeting and courageous in their fight against the focused attention of the British fighters.

*20:47 hours*

The hostile blocks moved south. Winston sat slumped in his chair, feeling the tears welling in his eyes. He could only imagine what the young fighter pilots experienced flying into formations of five to ten times their numbers.

"It should be over for the day, Prime Minister," said Park.

Winston did not respond at first. He wanted to shake the hand of every pilot, as he struggled with how to recognize such heroic, mortal and determined combat. He wiped his eyes with a handkerchief before answering Park. "We have witnessed a most heroic battle."

"Yes, sir, but unfortunately I believe it is only the beginning."

"Nonetheless, this day shall live forever in history. The courage and skill of your pilots rank amongst the highest of all the warriors in human history. You are to be congratulated Marshal Park."

"Thank you, sir, but the battle is not yet won."

"Accept your just honor. No matter the eventual outcome of this great conflict, the honor of this day shall not be diminished."

"Yes, sir," answered Air Vice-Marshal Park.

General Ismay waited with Thompson and the driver at the automobile. The tail end of evening twilight gently lit the high, wispy clouds in the western sky as they departed Uxbridge for the return to Whitehall. Winston could barely contain his emotions as he repeatedly placed himself in one of those small fighter cockpits high above Southern England. Listening to the desperate sounds of battle and observing the display of the tremendous numbers involved in the day's action made the images even more dramatic. To Winston Churchill, he and a few others had borne witness to the purest demonstration of courage under fire he could recall in all of history. The rarest of moments captured Winston's imagination and strengthened his resolve. He could never allow himself or anyone else to fail these dauntless young warriors.

"A most impressive exhibition, I must say," General Ismay said, trying to draw the Prime Minister out of his brooding thoughts.

"Don't speak to me," responded Winston, far more sharply than he really intended. He churned through the stream of images and thoughts in his head. He had not meant to bark at Ismay, but the intensity of his thoughts overpowered his reason. Winston eventually decided to make amends. "I have never been so moved, 'Pug.'" Ismay nodded his head in agreement. "I have tried to recall my history as best I am able," Winston paused to rub his tired eyes. "I do believe we have witnessed a truly historic day, possibly without equal in the chronicles of human struggle.  The only episode that comes to mind is the valiant stand by the Spartans at Thermopylae."

General Ismay nodded his head again as Churchill slumped in his car seat, lost in his thoughts of history and the significance of the dramatic day. Winston knew quite well that Air Vice-Marshal Park was precisely correct . . . the battle was far from over, but the country needed a few heroes to focus their spirits, their toil, their struggles.  The young pilots in the Spitfires and Hurricanes were truly those heroes, godsent at a most critical juncture in the Battle of Britain.

"General," Winston said, as he turned to his riding companion with tears in his eyes, "I do believe we will find that no one has ever owed so much to so few.  We do not yet know the full extent of the damage, but the numbers those pilots faced were overwhelming . . . it staggers the imagination."

General Ismay saw the emotion in the Prime Minister's face, as the dim light from the blackout-slitted headlights of an occasional passing automobile flickered across his round, stern face.  The Prime Minister made no attempt to hide the profound impact the day's events had on him personally.  That unspoken candor had its own impact on the seasoned Army general who was asked to relieve the accomplished and distinguished Army leader, General Lord Gort, VC, GCB, CBE, DSO, MVO, MC, as commander of the beleaguered and retreating British Expeditionary Force in France during the miracle of Dunkirk.  Even the general swallowed hard to hold back his emotions.

"I must tell the nation, the Empire and the world what has happened here today," said Winston Churchill, as both men gathered themselves and they made their way back to Whitehall at a far less frenetic pace.

"Without question, sir.  I am certain it will be as uplifting to the people as it was to us."

"I sincerely hope so.  We need the light of hope to see us through this storm."

They rode the remainder of the journey back to Whitehall and No.10 Downing Street lost in their thoughts.  Winston could not avoid the recognition that London was not the same city without its lights.  The people seemed to be going about their nocturnal activities, but it simply did not have the same gaiety as it did before the threat of nighttime bombing extinguished the lights.  He wanted the great city to return to its natural state soon.  Whitehall was empty as they turned past the armed, attentive and dutiful Tommies onto Downing Street.  Winston thanked General Ismay for his company and left him for his ride home.

———

*Thursday, 15.August.1940*
*No.10 Downing Street*
*Whitehall, London, England*
*21:50 hours*

John Martin, Churchill's trusted and loyal Principal Private Secretary, waited for the Prime Minister's return. He stood as Churchill entered the office.

"John, you certainly did not have to wait for me," Winston said.

"Three messages of importance, sir," said Martin. "Sir Edward received for you a message from 'Intrepid' as well as the preliminary results from today's air operations. Also, Mister Chamberlain rang earlier for you. He would like you to call this evening regardless of the time. I must say, sir, the doctors tell me he is terribly ill. He is in his final days, they say. The cancer is not treatable."

Winston took the sealed envelope from John Martin. "I might tell you, John . . . General Ismay and I witnessed what I suspect is the most heroic endeavor this afternoon. The flood of Nazi bombers was turned back by such a few fearless, courageous and selfless fighter pilots. While the battle is still joined and we are far from salvation, the world shall know what these young men have done for freedom."

"It was that good?" asked Martin.

"It was the most profound and emotional moment in my lifetime, and that is saying quite a bit." They laughed at the many trials and tribulations of Winston Spencer Churchill in his 66 years of exploits in India, Sudan, Cuba, South Africa, France and Parliament. John Martin knew as well as anyone there were few people who could rival Churchill for lifetime accomplishment. It was the experience of Lord Randolph's son that added the real poignancy to Winston's words of description. "Now, you must go home. It is late, and we have many arduous days ahead."

"Are you sure, sir?"

"As sure as I can be about anything these days."

"Very well, then I shall beg your leave."

"See you tomorrow, John."

"Don't forget to call Mister Chamberlain."

"Thank you, John. Good night."

Winston Churchill sat at his desk with only the small desk lamp for illumination. He stared at Bill Stephenson's envelope and thought of the other things before him. How would he tell the people what happened on this day? How could he find the words to do justice to the enormous exertion? Could they be at the watershed for the battle? Had they turned the corner?

He looked at the single page, typed note from Sir Edward. The preliminary count from the Air Ministry told him the Germans lost 75 aircraft against 34 for the RAF. Eighteen British pilots killed and not quite double that number wounded. The size of the German Air Force meant they would win a war of attrition in the air. Winston remembered clearly the charts presented by Sir Hugh Dowding at that crucial War Cabinet meeting last May that decided the British reinforcements during the Battle of France. The RAF could not afford the horrific loss rate for long. The Battle of Britain had to be won soon, or a tragic result would be inevitable. Winston turned his attention to Stephenson's message. He opened the envelope. There were two sheets of paper, one from 'Intrepid' and the other from 'POTUS.'

---

## MOST SECRET

```
15th August 1940
FROM:  INTREPID
TO:  PM
     Attached please find a personal note
from POTUS that he asked me to deliver
privately.  My last discussion with him was
most productive.  He remains in high spirits
and optimistic about the future.
     As you probably know, WHIZZBANG arrived
safely.  Counterparts appear to be ecstatic
over the cornucopia of items to be delivered
early next month.  More formal assessment from
whizzbang to follow.
     Discussions with big bill and others
continue to progress quite well.  Hope to have
agreement on destroyers soon.  Counterpart
proposals positive in light of national mood
here.  Expect successful conclusion within
fortnight.
     News of monumental air battle commands
considerable attention here from every level
of society.  USGovt has allowed NY Times to
```

```
publish bold advertisement for pilots to join
RAF.  This is a subtle but very big step for
POTUS and USGovt.
     We all wish we could be of more
assistance to you and RAF in this valiant
struggle.  Our prayers, hopes and best wishes
are with you and the nation through this trial.
```

## MOST SECRET

---

Winston Churchill smiled as he received the warmth of another ray of hope.  He sensed events were beginning to turn their way after so many months of disappointment, disaster and depressing news.  As Stephenson said, the subtle movement of the President as well as the Government of the United States of America was probably as significant as the events of the day, but could not be publicized or acknowledged in any overt way that might jeopardize the progress.  Winston knew the United Kingdom and her Commonwealth could not easily win the war without the Americans, which is why Bill Stephenson's effort to garner the necessary support was so critical.  It simply had to be.  The principal question was, when?  Could the Fighter Command hold sufficient strength in the air to prohibit an invasion this summer?  Could the British exert enough pressure to keep the Nazis occupied until the Americans did enter the war?  What if the Japanese decided to take an even more aggressive stand in the Pacific?  What would the Soviets do through all this?  The questions came faster than answers or even guesses.  Maybe President Roosevelt's note would have some more good news.

---

## TOP SECRET

```
58
August 15, 1940
FROM:  POTUS
TO:  Former Naval Person
     We listen with keen anticipation to the
progress of your current aerial battle.  I must
tell you our ambassador believes the defense
of Great Britain is lost and the overwhelming
strength of the Germans cannot be withstood for
very long.  While his messages are troubling
to me, I find considerable reason for optimism
```

```
listening to the broadcasts of our cbs news
correspondent, Edward R. Murrow, on assignment
in London regarding the heroic battle overhead.
The nation listens and watches with genuine
concern and anxiety.  I truly must believe the
cause of freedom shall be borne on the wings of
your gallant aviators.
        Negotiators make good progress toward
50 destroyers deal you have requested.  I
believe this arrangement can withstand adverse
reaction in Congress.  I sense the mood of the
nation shifting ever so slightly toward at
least indirect support of British people.  I
still cannot see active military support on
the horizon although the sympathies of our
armed forces are clearly with you.  It is
public opinion and congressional antipathy that
concerns me the most.
        Our prayers are with you and your pilots.
As always, godspeed and following winds, as you
go in harm's way.
```

**TOP SECRET**

---

Tears rolled down Winston's cheeks once again.  The words from Franklin Roosevelt added to his growing optimism and fueled his need to convey the bright side of the present struggle.  He looked at the wall clock in the dim light of the Prime Minister's office.  It was approaching midnight and he still needed to call Neville.  Although the message said no matter what time, a terminally ill man did not need the extra stress of a late night call, however, Winston knew he would appreciate the news.

Winston lifted the handset and buzzed the switchboard.  The perky female voice took his request, and performed her duties promptly and precisely.  "Go ahead, sir," she said, once the connection was made.

"Neville, this is Winston," he began.  "I do apologize for the late hour, but it has been a rather momentous day."

"Not to worry, Winston.  I needed to hear the news from you.  I knew you would be in the thick of it, and I heard the bombs along with that dreadful pulsating droning of those engines."

"As you surmised, I did manage to make it out to Uxbridge for the last half of the day's events. You would be proud, Neville, as the nation will be, of our dauntless young lads throwing themselves into such overwhelming numbers of Nazi bombers and fighters. The German losses were three to one," Winston said, not minding the slight exaggeration. "So few of them and they are our principal bulwark against the invasion. They stood the most onerous test today, and they have lifted my spirits."

"You shall talk to the nation?" Chamberlain asked.

"Yes, most definitely. The pilots deserve the recognition, gratitude and praise of the nation. We need to feel hope they offer us in this rather dark period in our history."

"I am not sure how much more time I have before I shall meet our Maker's judgment, but there are several thoughts that I could never forgive myself for passing without saying. The nation owes you an incalculable debt of gratitude, Winston."

"Nonsense."

"I must ask you to let me speak. So, if you please, let a dying man speak his mind and voice his confession," he said and paused partly to take a deep breath, but also to let Winston acknowledge his request. Churchill remained silent in deference to his esteemed colleague of many years. "I have never told you this, Winston, but I want you to know I have always respected your courage, honor and persistence in the face of such formidable resistance. All those years of adversarial rhetoric, and it was you who was right all the while. I offer you my humblest and most heartfelt apologies for the trauma I have caused you and your family. It was never personal, and I trust you will forgive me."

Tears seemed to be endless for Winston this day. The frail old man of the Conservative Party had always had the best intentions and motives at heart. It was his counterpart in the negotiations of 1938 that had been devious, deceitful and villainous in his heart. The former Prime Minister and leader of the Conservative Party, and now Lord President of the Council, the Right Honorable Neville Chamberlain, MP, remained a man of honor with the most lofty of intentions and desires – peace on earth. He had endeavored to hold the peace against exhausting waves of tyrannical megalomania.

"Thank you, Neville."

"No, thank you, Winston, for being there when I needed you, and for being where you are at this most critical time in our nation's history."

"Thank you, again."

"Now, tell me of the day's exploits," Chamberlain said with child-like enthusiasm.

Winston Churchill took another quarter of an hour to convey the facts, as he knew them, as well as the emotions he felt observing the conduct of the great air battle.  Chamberlain seemed uncharacteristically interested in the conduct of the war and especially the efforts of the RAF fighter pilots.  When they concluded their telephone conversation, Neville Chamberlain repeated his apology, thanked the Prime Minister for the courtesy of a personal briefing and wished Winston Churchill the best of luck on the inexorable march to victory.  Winston sat there with the thought . . . will wonders ever cease?

———

*Friday, 16.August.1940*
*RAF Middle Wallop*
*Middle Wallop, Hampshire, England*

**P**ilot Officers Brian Drummond and Jonathan Kensington walked casually along the western boundary of the airfield.  Both pilots had eaten their breakfast and left the Officer's Mess earlier than the others.  The damp, heavy morning air left sparkling droplets of dew like diamonds laid out before them.  The early morning sun and hazy, humid sky foretold a carbon copy of the previous two days.  The chirping, singing birds along with the occasional groan of a distant cow brought an idyllic mood to the rich, earthy smell of the land around them.  Birds fluttered and a hedgehog ran out from the adjacent hedgerow across the open grass of the landing area.

"Why do they get us up so early when the Germans don't start their activities until mid-morning?" asked Brian, as they reached the halfway point to their new Dispersal tent.

"The Sun."

"They could at least give us a little more sleep."

"Why?  Nearly everyone goes back to sleep at Dispersal."

"Wouldn't it be more comfortable?"

"Everyone is so tired, Brian, do you really think it matters?  They will sleep in the cockpit given half a chance."

"Some already have," Brian said, as they laughed at the thought.

An Air Force truck passed by them on the other side of the hedgerow going south on the A343.  'Boxer' Stockard and 'Red' Burns were standing in the back holding onto a cover railing.  From their vantage point, they could see over the hedgerow.

"You guys needin' a little exercise," shouted Burns.

Brian and Jonathan both waved, as if they were acknowledging their adoring fans. The truck carried most of the pilots between the Officer's Mess and the Dispersal tent. This particular morning, Brian wanted to feel a little bit of normal walking in the country, as he often did in Kansas when he wanted to think. As was usually the case, he asked his best friend to join him for the mile or so walk.

"How is Rosemary?" asked Brian, as they watched the No.609 Squadron pilots take their places around the tent a quarter of a mile ahead of them.

"I was wondering when you would get around to asking."

"Sorry, Jonathan. Anne's demise has been pretty heavy on me, and then having my life saved by a rather attractive . . ."

"And older."

". . . farm woman has really mixed things up."

"Since you asked, Rosemary as been asked to stay at Carlingon until the invasion threat subsides. Strange thing is, part of her wants to be here, near you, for God's sake." They laughed with Jonathan's sarcasm. "You are going to break her heart, Brian," he said more solemnly.

"Not intentionally."

"No, probably not but you will nonetheless."

"I like Rosemary a lot, and it is so confusing. It's probably quite stupid and immature, but I feel a kinda mystical sort of connection to Charlotte Palmer."

"She did save your bleedin' life, after all, you sod."

"Maybe that's it. Maybe we are kindred spirits, as they say."

Jonathan Kensington chuckled and waved his hands, as if he were waiting for a large ball to be thrown to him. "Brian, a word of advice from a slightly older friend, don't take this too far. She is an ordinary woman with her own problems and her own baggage. Please don't romanticize her actions as a citizen trying to help a warrior in trouble. It doesn't work that way."

Brian thought about Jonathan's admonition. He was probably correct, Brian told himself, but the feelings were there nonetheless. "Maybe not," he answered pensively, as he tried to meander his way through the rocks and shoals. "I would still like to know her better."

Jonathan laughed. "Know her better. You mean know her better in the biblical sense, you bleedin' romantic old twit."

They both laughed and jabbed at one another like schoolboys. The others watched, waited for them and provided appropriate comment as the last two of their number reached the remote dispersal tent. The small band of dauntless aviators turned their attention to the east as the initial reports of morning combat made their way to the large tent at the southwest corner of RAF Middle Wallop.

It was mid-morning with sounds of a rural country farm all around them that made the news seem so surreal. Only the sleek, graceful curves of the now ten Vickers-Supermarine Spitfire Mark IA fighters scattered in front of them brought a focus to reality. The news from Kent and Sussex continued to build through the morning, and yet No.609 Squadron remained at Available status. The weather was near perfect. There was an abundance of plump targets in Hampshire and Dorset, but they remained . . . sitting . . . waiting . . . contemplating what lay ahead. When would the call come? When would the enemy's attention shift to the West?

The interminable cheese sandwiches and tea for a light mid-day meal added to the mounting frustration. Several of the pilots wanted to perform a check flight on their aircraft just to do something and break up the agony of waiting.

Squadron Leader Darling came around the northside of the dispersal tent. "Well, gents, I have a piece of news for you."

"We've been released," interjected Flight Lieutenant Morrow.

"Unfortunately, I'm afraid not. But, I think it only fair to inform you I have just concluded a business transaction with the farmer who owns the dwelling across the carriageway from us."

"What sorta deal, Skipper?" asked Flight Lieutenant Beamish with his best, distinctive Scottish brogue.

"I have convinced the gentleman to let a small cottage behind his home. My wife, Emily, has insisted upon joining us. We shall be moving her into the cottage this weekend."

"Isn't that a bit close for a loved one, Skipper?"

"This was her idea and as some of you know she can be quite insistent as well as persistent. She says she wants to bake us some good biscuits to go with our afternoon tea."

"When we get afternoon tea," grumbled Pilot Officer Stockard.

"My recent wounds have apparently precipitated this rather impulsive action by my wife. I sincerely hope she will not interfere with our duties or camaraderie."

"She'll be good company for me while you blokes are off defending the faith," said Corporal Warren in a somewhat rare participation.

"Indeed."

"There you have it. It's a done deal then."

"She'll be one of the boys," said Pilot Officer Burns.

"Dear God, I hope not," answered Darling to the resounding laughter of the group.

As the afternoon ground on and they remained at Available status, even Squadron Leader Darling became somewhat irritated with the waiting. He called No.10 Group Headquarters in Box, Wiltshire, as well as Middle Wallop Sector Control several times partly to make sure there were no problems with the telephone lines as well as to ensure something was not missing. Each time he was assured the communications system was fully functional. The waiting was made all the more frustrating as the reports and information coming to them described the significant assaults on the No.11 Group fighter bases.

*18:17 hours*

The merciful end came with their release from duty for the day. Squadron Leader Darling walked across the A343 to check on the cottage Emily Darling would soon occupy. The others including Brian and Jonathan took the truck back to the Officer's Mess. The evening meal passed quickly as Brian and his colleagues compared stories with the pilots from the other squadrons. The night fighter pilots of No.604 Squadron were just getting ready for their duty period.

The mood struck most of the Middle Wallop pilots as they decided tonight was the perfect occasion for a good old-fashioned pub-crawl. They worked their way toward Andover and the Black Swan pub with its magnetic attraction of rather fun-loving young women who enjoyed the carefree, fearless, and maybe a touch crazy, young men who flew the curvy airplanes. Brian resisted initially but succumbed to the temptations, as they wandered up the A343 stopping at nearly every pub along the way. They were in rare form by the time they reached the Black Swan and the clientele was in an appropriately encouraging mood.

They heard from several other pilots about the extraordinary actions of Flying Officer James Nicholson going down in his burning Hurricane, and as he was struggling to bail out found a Bf110 directly lined up in his sights. He stayed in his mortally wounded fighter to squeeze off several bursts downing the German twin-engine fighter before finally bailing out with burns and other injuries. Nicholson's exploits warranted several long drinks and bravado throughout the establishment. They also heard about the fake landing by two Ju88 bombers at RAF Brize Norton, the primary pilot training base in Oxfordshire where Brian started his RAF training. They approached with the undercarriages down to look like two Blenheim aircraft, and then making it over the field, they raised their wheels and dropped their entire bomb load on the lines of Tiger Moths, Gladiators, Tudors and other training aircraft.

The attack had been unchallenged and devastating to the important training base. They drank some more to future opportunities to even the score. The talk was strong, and the citizenry and especially the young women absorbed and reflected the bravado.

The night blurred. Most did not hear, recognize or care about the sounds of twin-engine Bristol Blenheim Mark IF night fighters taking off climbing into the black sky toward an invisible target high above them. Pilot Officer Brian Drummond was no exception.

———

*Saturday, 17.August.1940*
*Standing Oak Farm*
*Winchester, Hampshire, England*

For the widow, Charlotte Palmer, the late morning telephone call from Pilot Officer Brian Drummond, the young American volunteer pilot she rescued from her pond, seemed both appropriate and yet quite odd. Initially, she did not think his requested visit was a good idea. She had no gentlemen callers at her farm even after her husband, Lieutenant Ian Palmer, Royal Navy, was reported killed in action aboard HMS *Glorious*. Her routine to manage the affairs and business of the farm occupied her days and often much of her nights as well. The fact that she had given her helpers the weekend off after nearly three straight weeks of work added to the oddity. She would be alone. What would she do? She hardly knew much about the young man with a strange but seductive accent and manner of speaking.

Charlotte remembered the events of that afternoon 2½ weeks previous when she rescued the pilot from drowning in her pond. The horror of seeing the explosion, and then the white blossomed semi-sphere of his parachute as he descended unconscious into her pond continued to give her shivers. Meeting him at the Winchester Hospital made the intrigue and fascination all the more sharp to her. He was grateful for her initiative and commitment; that she appreciated, but it was his confident, boyish brashness that struck the chord with her. He had not shown the slightest concern for his own safety and the incredible close brush he had with death. The young man displayed no signs of being shaken by his experience.

As she waited on lunch for his arrival, Charlotte found her other thoughts. He was tall and projected a larger presence. His angled, well-defined facial features matched perfectly with his wavy-brown hair and those seductive blue-gray eyes that seemed to suck her into their grip. Ian's passing less than two months ago reflected a harsh, bright light on this young man's visit. It was far too soon for a gentleman caller.

The rigors of managing a farm helped ease the pain of her loss. Charlotte did not know whether it was just fatigue, masking the pain like morphine, or her wounds were truly healing rapidly. The discomfort of this confident, a bit cocky, youthful RAF aviator stirred up many more questions than she had been able to keep hidden. Charlotte knew she had most of her life still ahead of her, but the close, intimate companionship and connection she lost with Ian created a void that she also recognized had to be filled someday. This was simply too soon.

Charlotte Palmer used the remaining minutes until he was due to arrive to fret about the arrangement of pillows on her divan, the several bouquets of fresh cut flowers and the cleanliness of the table tops. She felt younger with her anticipation. After all, he was quite attractive, a hero some said, a bit exotic and an American.

The crunch of the gravel in the driveway surrounding the engine sounds of the motorcar announced his arrival. Charlotte found herself fidgeting and dancing like a schoolgirl. She took a few deep breaths to calm herself. The Hampshire native walked to the heavy, dark, oak door with ancient cast iron straps and hinges, and opened it with surprising ease to see the stunning RAF officer extract himself from the diminutive Morgan.

"Good afternoon, Mister Drummond."

"And good afternoon to you, Mrs. Palmer."

"I trust your journey was uneventful."

"Quite."

"Excellent. Would you care to come in, and would you like some lunch?"

"I don't want to be a bother, ma'am."

Charlotte smiled. "First, it is no bother. It is just a simple country midday meal. Second, please don't call me ma'am. It makes me sound so old. I would prefer Charlotte, if you please."

"As you wish," Brian responded with a broad smile.

She motioned for him to sit at the dining room table in a small room adjacent to the kitchen. She talked about the war, as she perceived it, while she completed the final preparation of the bread, crackers, cheeses and sausages. Charlotte poured a glass of red wine without thinking to ask him if that was what he wanted. She fought to control the nervousness she felt and that effort transformed into constant motion and talk. As the wine began to calm her, the path of the rambling conversation began to narrow toward the events of just over a fortnight earlier.

"I know this is a little awkward given the bad luck of HMS *Glorious*, but I felt it was extremely important for me to come to you, to thank you in any way I can for what you did."

"Nonsense. I did what had to be done."

"Maybe so," Brian said calmly, "but . . . well . . . ," he paused, as he struggled with the words he wanted to say, "well, I feel a connection to you." The surprise in her eyes and the stiffness that sprang through her body communicated her apprehension, and maybe even fear. "Sorry, ma'am . . . er . . . Charlotte. I didn't mean to offend you. It's just that you saved my life. You gave me my life back."

The stiffness kept her distant. "I accept your gratitude, Mister Drummond."

He held up his left hand. "Please, I must insist you call me Brian."

Charlotte laughed and began to feel the tension ebb. She nodded her consent.

"This may be hard for you to understand, but doing what we do, we live on the edge of a very steep cliff. I've been shot down several times, but for some reason, I have never felt I was going to die. It probably stems from a crash I had many years ago when the plane I was flying hit a dog and cartwheeled over the ground. My teacher threw me back into the air, and it has been that way ever sense." She nodded but let him continue. "I don't remember what happened this last time. I only have the information you provided, but, well, the best way I know to describe my feelings is, well, it is kind of spiritual, maybe even religious, like you're my angel."

Charlotte chuckled and pushed the breadcrumbs around her plate. "I can assure you, Brian, I am no angel. I am just a simple woman trying to hold things together. You happened to land in my pond. I could not very well stand there and watch you drown."

"From what my mates told me, you had to fight pretty hard to get me out of the water, and the doctors and nurses told me you breathed life back into my lungs."

Charlotte Palmer felt the flush of embarrassment wash over her. His description seemed more intimate than she remembered, but the essence was correct. "As I said, Brian, I simply did what had to be done."

"I can never repay you for your courage and extraordinary efforts. I owe you my life."

Charlotte instantly felt an enormous weight on her shoulders. Was he saying she was now responsible for his life? Was he expecting to do something for her? What did all this mean? Charlotte Palmer was getting uncomfortable with this strange sense of responsibility his words implied, as if she was now in command of his life. She needed to change the subject. "Would you like to see my farm?" she asked.

"Certainly."

She led him out of the cottage and began slowly walking.  She described the various features of her farm from the large wooden barn to the small, stone, food storage building between the cottage and barn.  Although he did not say he lived on a farm in America, the young flyer knew more than any city person would about the animals, equipment and the farming process.  They laughed and shared experiences with animals as well as the importance and beauty of nature all around them.  She was fascinated by his tales of massive, earth shaking thunderstorms and even more so by his description of tornadoes.  She and Ian had seen the Wizard of Oz in London before the war, and those unusual storms possessed an odd attraction.

Charlotte, with Brian's urging, reenacted his rescue without jumping into the pond.  The words seemed so serious at first as she described the scene in the sky above her as she rested at the crest of the small hill.  Brian managed to laugh and joke about the events that afternoon which infected Charlotte with nearly uncontrollable laughter.  They enjoyed the scenery; the warm summer air and the light breeze that kept them reasonably cool.

The Sun was still a couple of long hands above the horizon when she said, "I'm afraid, Brian, I have farm duties I must tend to."

"I'm not a stranger to the farm.  I can help you."

"It is that time of the day, and I could use some help.  Have you milked a cow before?"

"Well, maybe I haven't done everything, but I'm a quick learner."

"I'll bet you are," she said, as she walked toward the barn.

Charlotte Palmer had four cows that waited at the barn door groaning in protest of their distended udders.  She opened the door allowing the cows to walk to their respective feeding slots.  She showed Brian how they harnessed the cows to hold them in place.  The stool and tall bucket came from the other side of the barn.  He removed his hat and coat, loosened his necktie and rolled up his shirtsleeves to his elbows.

"Here you go then," she began.  "Always sit on the left side of the cow."

"Why?"

"I don't know, just do.  Stroke her side a little to let her know you're there, and you're not going to hurt her.  Sit yourself down," she said, stroking the cow's side and patting the stool.  "First, you should always massage her udder to feel how full she is, as well as help her milk to let down.  Sort of a natural action, I suppose."

Brian reached for the cow's udder and touched it quite gently.

"You can do better than that," she said reaching past him caressing the large mammary gland firmly. She felt her left breast brush across his arm and tried not to react. "Firmly. Feel what she has." Charlotte watched him and saw the cow getting a bit fidgety. "Now, grasp a couple of her teats at the base, squeeze firmly and pull down pointing the stream into the bucket." She watched him make several pulls, and then knelt beside him placing her hands over his. "Reach up a little higher on her teat to the base of the udder. Squeeze her like you would . . . ah . . . well . . . firmly." She stood. "I think you got it." Charlotte retrieved her own stool and bucket.

They worked without words. Charlotte chastised herself for carrying on with him, but it just seemed to flow from her. The attraction was undeniable, but she knew she could not let herself be swayed. She finished her first cow before Brian finished his first cow. She poured her bucket into a large storage container on the other side of the barn.

"How do you know when you're done?"

"When no more milk comes out," she answered curtly.

He kept at it for several minutes. "This one's done."

"Pour the milk into the vat over there," Charlotte said, as she pointed toward the container.

She watched him carry the bucket and pour. He had a strong, athletic character to his movements, so fluid and yet so precise. His size was deceptive, masked by the grace with which he moved about the barn.

"You are welcome to milk the fourth, if you care to," she said calmly, as she worked on the third cow.

"Excellent," he said trying to replicate a London accent.

He moved to the fourth cow. Charlotte looked under the belly of her cow to observe his approach and initiation of the milking process. If she had not known better, she would have assessed him as an accomplished practitioner. His hands seemed to enjoy the cow's udder, and the cow made the strangest sound, almost a sigh, as if she derived pleasure from his strokes. Contentment filled the large, round, dark eyes of the cow as she looked back to see for herself what was happening. Charlotte Palmer was amazed and fascinated.

"I think I'm getting the hang of this," he said over his shoulder, not really looking anywhere.

Charlotte sat upright, stopped her milking and froze for a moment. This whole situation was beyond real. Something had to be wrong. She pinched her forearm as if to see if she were dreaming.

"Are you there, Charlotte?"

She cleared her throat, told herself to sound preoccupied, and said, "Yes. Nearly finished with mine, I should think."

The rhythmic squirts of milk, one set into a bucket full of milk and the other echoing into the metallic receptacle, filled the barn. Several finches alighted in the rafters and began a serenade.

"I think I rather like this milking stuff," he said.

I bet you do, you nasty little bugger, she said to herself. The lewd but pleasurable thoughts brought a flush of embarrassment she did not expect. Why was she acting like a damn schoolgirl? He was just a man, a younger man, whom she rescued from her pond.

Charlotte finished her effort and completed the remainder of the afternoon ritual. She could not release the other cows, as that would only cause problems with Brian's charge. She stood behind him. He had strong, maybe even very strong hands, clearly evident as she watched him. They talked about the farm, cows, goats and the other animals as well as her ample garden. He did understand farming. His uplifting, nearly laughing delivery of his words made the troubles of the world beyond disappear.

They finished, released the cows, cleaned the buckets and barn, and returned the equipment to its proper place. They laughed and joked about silly things.

She motioned toward the door expecting him to retrieve his coat and hat. In a flash, he put his arm firmly around her waist drawing her around and to his chest. He kissed her delicately but with a certain passion. She hesitated for just a moment, partly surprised, partly joined to him, and then she pushed away and slapped him across the face very hard. Then, it was he who was shocked.

"You must go now, Mister Drummond," she spat. His expression turned instantly from the shock of her reaction to profound apology. "Now," she shouted.

"I'm sorry, Charlotte."

"No!"

"I don't know what came over me. I'm so terribly sorry," he said with considerable sincerity.

"I shall not say it again. I want you to leave immediately," she said with force, and an arm raised straight and her first finger pointing toward his car.

He picked up his hat and coat, and left her, without another word, standing in the barn. The sounds of the car faded into the distance, before allowing herself to release the incredible tension wound up within her. She sank into a pile of hay and cried hard. Her body shook with the combination of release, regret and remembrance.

———

# Chapter 5

Only the consciousness of a purpose
that is mightier than any man and worthy of all men
can fortify and inspirit and compose the souls of men.

-- Walter Lippmann

*Sunday, 18.August.1940*
*RAF Middle Wallop*
*Middle Wallop, Hampshire, England*

## Week 7

The day started hot.  The Sun had barely risen above the eastern horizon when the wake-up call pronounced the beginning of another day.  The humidity laid a heavy haze across the morning Hampshire countryside.  Brian walked with Jonathan and one of his fellow Americans, Frank Burns, to the truck that would carry them to their dispersal tent.  Brian wanted to discuss his unfortunate visit with Charlotte Palmer, but Jonathan was about the only man he felt comfortable with regarding personal issues.  Burns, while a vital squadron mate, had not yet become a close friend.  The Ohio State University alumnus seemed to always make Brian feel younger, less experienced, less sophisticated and less important.  Brian felt no desire to perpetuate those feelings and did not seek his counsel or company.  Jonathan, on the other hand, even as a graduate with honors from Cambridge University, never made Brian feel less worthy.  Brian did not dislike Frank Burns, he just did not like being around him other than in the air.  In the skies, Frank was a skillful and accomplished fighter pilot.  Brian respected Frank in the air, and the reciprocal was also true.

The truck ride to the southwest corner dispersal point was hardly worth the effort although all the pilots displayed various signs of fatigue.  They hardly said anything to Brian as he joined the group for morning alert even though he technically was still on his rest pass until noon.  Brian felt the need to fly, to put his worldly cares behind him.

Each pilot took a chair since the grass was still damp with dew and drifted off into their thoughts.  Brian was no different.  For some odd reason, he began to take an accounting of the last two extraordinary years of his young life.  The jump from the simple life on the Great Plains of Kansas to a Royal Air Force Spitfire fighter pilot in one of the southernmost counties in England made him think of Dorothy, the Yellow Brick Road, the Wizard of Oz, and 'We're not in Kansas anymore, Toto.'  Somehow, the whirlwind lifted him from a safe future to the thrill, exhilaration and danger of the moment.  His thoughts could not avoid Jeremy Morrison and especially Anne Booth.  He refused to acknowledge the dark side of what happened and chose to focus on the delightfully enthralling, good times.  In his mind, Anne remained the good person she had always been to him.  Rosemary Kensington and Mary Spencer brought a smile to his face as well.  Even the illicit facet of his relationship with Mary could be overlooked in the warmth of her gentle, if ever so slightly desperate, yearning.  However, it was Charlotte Palmer that truly intrigued him.  While each of the women he had known was strong and confident in her own way, it was Charlotte that commanded the majority of his attention.  Was it the newness, the rejection, the distance, the intimacy of her as savior?  Brian could not grab a hold of anything in particular.  It could be all of those and more.

"Mister Drummond," said Squadron Leader Darling instantly snapping him back to reality, "what are you doing here?  You are supposed to be resting."

"Nothing to do, Skipper."

The grumblings, stirrings and quips from the others enhanced the reproach of Darling's query.  The strange part was he really did not have anywhere else he wanted to go until he figured out how to apologize and return to the laughter he enjoyed with Charlotte.  The thought of sleep in his bed in the Officer's Mess with the prospect of a German bomber depositing its destructive ordnance load on him was more than he wanted to consider.

"He has girl problems, Skipper," said Pilot Officer Roland Stockard with his eyes still shut.

"No, I don't," barked Brian.

Everyone came to alertness.  "Ah, ha.  The fishing lure hooked a big one," said Flight Lieutenant Robert Morrow.

"Enough," Darling shouted.  "We may not get many more opportunities for rest.  In the future, if there are any more rest periods and you are not going to avail yourself of that time, I would enjoin you to relinquish your time to others who also need the rest."

"Yes, sir."

The quips added some levity to the admonition as well as solidifying the lesson for Brian.  He needed to control his emotions to avoid the unwanted attention of his colleagues.  His thoughts had to remain locked up.

"If you are not careful," Stockard said quietly, "you will end up like your millionaire countryman."

Brian turned to Roland Stockard, not understanding what he was trying to say.  The confused expression probably gave Stockard a clue.

"Billy Fiske."

"What do you mean?"

"Isn't he the motion picture producer?" asked 'Red' Burns.

"Yes, so they say," answered Stockard. "He was well placed in the Millionaire Squadron."

"Six Oh One?"

"Yes, indeed, Six Oh One, the County of London, Squadron, the so called, Millionaire Squadron," interjected Robert Morrow. "More millionaires than any squadron in the world."

The weak, almost rattling, ring of the field telephone terminated any words or movement. "Mister Darling," came the call from Corporal Warren inside the tent. At least, it was not a scramble command so early in the day.

Darling joined them. "Seems we have a high flyer, recce intruder. Green Section. 'Jackstay,' you take your lads and have a go at them."

"Aye, sir. Here we go then, laddies," said Beamish.

Pilot Officer Janus Kradilcek jumped to his feet and practically ran to his fighter. Beamish and Drummond took a more restrained pace. All three Green Section airplanes gracefully took to the air. The single reconnaissance aircraft was reported by Sector Control to be heading generally toward them, probably directly to Coventry, Manchester and Liverpool, at 35,000 feet.

As they continued their high performance climb, Brian checked his oxygen mask several times. This would be the highest altitude he had ever flown. He remembered the stories about the effects of high altitude flight – the cold, the intense brightness of the Sun, and the lack of atmospheric oxygen to sustain life.

The condensation trail of their high altitude target clearly and distinctly marked its position, course, and relative speed. It also told them it was a twin-engine, medium-size aircraft. Based on the reported altitude of this fellow, it had to be one of the new Ju86P airplanes specially configured for high altitude, reconnaissance operations. Brian remembered the intelligence briefing describing the two-stage, turbo-charged, Daimler-Benz engines with their tuned propellers and the plethora of intelligence collection equipment expected to be on the machine. Several other pilots had tried to intercept the high flyer, but so far, none had been successful.

The three Green Section Spitfires continued to climb in a wide V formation toward their projected intercept point. Passing through 27,300 feet, their contrails blossomed. Their prey would now see the hunters.

"Sorbo Green, my birds acting a bit sickly," said 'Jackstay' Beamish at 31,000 feet.

Brian felt his controls become softer, mushy they called it, requiring more frequent, small inputs to keep the aircraft level and climbing.

"See if you can eek out some more and tag the bastard," he said.

"Roger," Brian answered.

The German maintained his heading, altitude and speed, as if in defiance of the now two fighters headed directly toward a point in front of him. Maybe the German felt so invulnerable he failed to spot the fighters. At nearly 34,000 feet, 'Crazy' Kradilcek waggled his wings, as he fought with the machine, and then fell off saying he could no longer maintain the climb. Brian gently coaxed the Spitfire higher. The German bomber refitted as a spy plane kept its track. As he waited for the closure to continue, Brian shivered as the cold of high altitude flight began to bite into his muscles and stab at his joints.

The Spitfire groaned in the thin air and began to level off at 34,700 feet despite Brian's attempt to keep her going. He watched the German spy plane pass just over the top of him. The plane looked quite similar to the more common Ju88 except for the modified engine cowlings, the glass windows in the belly and the lack of any guns or other protuberances on the fuselage.

He ever so gently pulled the nose up, quickly aligned his gun sight with an appropriate aiming point and squeezed off a burst. The firing of his machine guns along with the slightly nose high attitude caused the fighter to shake, stall with the left wing dropping, and then the nose dropped sharply toward the earth below.

Brian worked his controls carefully, slowly and as precisely as he could as he tried to recover and make another firing attempt. He quickly realized the futility and began his descent, as he watched the German continue on his mission, as if nothing had happened. Brian felt anger at the effortless defiance of the intruder. An airplane about twice his size, unarmed and showing no signs of evasive maneuvering or other reactions could so easily fly past the premier fighter in the world.

'Hunter' rejoined 'Jackstay' and 'Crazy' at about 25,000 feet. They radioed their lack of success with the high altitude intercept. They were many miles north. The industrial city of Coventry lay just ahead of them as they turned south. The canopy and especially the thick, forward bulletproof windscreen fogged up as his cold aircraft descended into the warmer, moist air of the lower altitudes. As Brian struggled to keep his canopy transparencies clear of obscuring condensation, he took stock of the experience. The cold, sloppy controls and poor performance added to the frustration of the freezing or fogging of the canopy convinced Brian the RAF was not properly prepared for the demands of high altitude combat.

By the time they reached RAF Middle Wallop, the remainder of the squadron had launched on another mission. It had taken them more than an hour for their attempted intercept of the German reconnaissance aircraft. The ground crews scurried about refreshing the three fighters in anticipation of future work. Corporal Warren waited for the Green Section pilots at the entrance to the remote dispersal tent.

"Where is the Skipper and the rest of our blokes?" asked 'Jackstay' Beamish.

"Scramble, sir. They launched three quarters of an hour ago along with Two Three Eight Squadron. Big raid on Portland, apparently."

"Great," growled 'Crazy' Kradilcek.

"Now, now. There will be plenty to go around, I'm sure," said Beamish.

While they waited for the others to return, Flying Officer Royster and his two assistants quizzed Beamish, Kradilcek and Drummond with more avarice that usual. The recent encounter was the first for any of them with the German, high altitude, reconnaissance aircraft, and they wanted to glean as much valuable data from the pilots as possible. They seemed to be genuinely intrigued by the contrast of poor, high altitude, performance of the Spitfire Mark IA and the sense of invulnerability the German crews must have felt. The analogy they used matched the experience. It was as if the intruder knew exactly how long the retention chain was on the angry watchdog and stayed just out of reach.

The low, deep throated, hum of many Merlin engines grew rapidly, followed by the protesting popping and rumbling of the engines at idle as the squadron returned for rearming. They taxied rather fast to their respective parking spots.

"We'd better gather up our kit, mates. Looks like we'll be going again shortly," said Beamish.

All three pilots grabbed their flight equipment. They watched the Spitfires stop and the crews go to work. All the gun port tapes were missing, and all the aircraft returned. Darling waved for Beamish and his wingmen to join him at the tail of his 'PR-A' Spitfire as the others slowly gathered around him. They joked about Green Section missing all the fun with the large Portland raid.

"As you heard, lads, Sector Control wants us airborne as quickly as possible. You have a few minutes at most to relieve yourself, get a drink or what have you, then let us mount up, so we can get back into it."

"Aye, Skipper," said 'Sparky' Morrow.

The pilots did as they were requested, and then began filling the cockpits as the crew finished on the fighters. As the last crew signaled their completion, Darling gave the crank up hand signal, and engines began to fire off. All 12 Spitfires were airborne within a few minutes.

Two squadrons of Hurricanes along with No.609 Squadron waded into a raid on the Isle of Wight and Southampton of more than 100 enemy aircraft about evenly split between medium bombers and fighters. The aerial combat lasted until they were out of ammunition and nearly exhausted their fuel.

RAF Tangmere was the closest Royal Air Force facility. Eastleigh was the closest airfield that could provide fuel but not ammunition. As they landed at Tangmere, the circles of fresh dirt all across the landing area as well as several damaged buildings showed the severity of earlier attacks. Most of the pilots remained in their cockpits as different, resident ground crews feverishly replenished the No.609 Squadron Spitfires. The wafting, intoxicating fumes of high-octane gasoline being pumped into the large fuel tanks just ahead of the cockpit mixed with the metallic sounds of ammunition trays being loaded and guns being prepared. The coveralls of the ground crew were dark with oil, grease and gun soot, but more significantly with large roundish patches of sweat. It was hot and humid. The crewmen gained very little evaporative cooling. Many stripped off the torso portion, wrapping the sleeves around their waists to hold the bottom portion up.

"Can I get you anything, sir?" asked a young leading aircraftman.

"I'm OK," Brian answered.

"American, ay?" smiled the man.

"Yes."

The crew chief extended his right hand, which Brian shook. "Good to have you with us, sir. We'll have you back up there shortly," he said, and disappeared before Brian could say anything.

As the crews began to finish, Brian looked across several aircraft in the staggered line to the 'PR-R' Spitfire. 'Slim' Koenig was bent over with his left hand on the tail supporting himself as he retched. Brian shook his head emphatically and wondered how long the older American had been dealing with his physical reaction to impending combat. Brian also could not avoid the natural question – how long could he keep going like this?

An officer ran from the Operations building to Darling's 'PR-A' Spitfire. In an instant, he was signaling the start up. Brian glanced over to Koenig who was scrambling back into his cockpit.

As they climbed at high power, Brian armed his guns and checked his machine to be ready for instant combat. A dull ache enveloped his body. Fatigue began its corrosive action on every muscle, joint and brain cell. Even his eyeballs ached. The dry, life-giving oxygen pumping through his mask made his throat sore. His stomach growled and rumbled violently from lack of food and the consuming demands of adrenaline.  How much longer could they keep this up?

They joined the fight already in progress over Southampton.  It was not clear to Brian whether it was the same group or others.  It had to be a fresh raid, but there was no way to tell other than time.  The short serpentine contrails of the fighters arced over the bombers.  The ferocity of the fighter engagements drained each of them further although none of them could recognize or acknowledge the sensations.  The white contrails soon capped the ugly black smoke of aircraft falling to the sea below.  Several Spitfires along with Bf109s stopped flying.  Bombers produced far larger smudges in the sky as the Hurricanes feasted on their slower prey.

A new raid appeared to the south over the top of their predecessors withdrawing toward occupied France.  Other squadrons were reported en route, but No.609 'Sorbo' Squadron positioned for the attack on the new raid.  They were low on fuel and ammunition, which meant they would have to disengage before the mission was complete.  They would soon be more vulnerable.  No.152 'Maida' Squadron would need 20 minutes to takeoff and climb to the engagement altitude, if they were not already in the air.  Brian glanced at his fuel gauge.  He had barely 20 minutes of fuel remaining at the emergency power settings they demanded for combat.  'Maida' had to arrive sooner.

They waded into the German escort fighters and were soon immersed in the whirlpool of modern fighter combat.  Brian was not the first, but he soon joined the growing number of his brethren who ran out of ammunition.  'Spike' Darling ordered the disengagement of No.609 Squadron.  Brian found an opening, rolled his machine on its back, pulled the nose down sharply, and began a maximum speed dive toward the coast while he swiveled his head looking for attackers.  He caught glimpses of the others doing the same.

Satisfied he had no tails, Brian took a broad view of the sky.  The distinctive shape of the No.152 Squadron Spitfires drew his attention, as they jumped into the fight with the Germans well above them.  The explosion of bombs all across the Solent leading to Southampton indicated the success of the Hurricanes in rattling the German bomber crews who decided it was better to drop their loads early and run rather than submit themselves to the deadly gunfire of the British fighters.

They had sufficient fuel without much margin for the return flight to RAF Middle Wallop.  As the adrenaline began to wear off, the numbing, aching weight of exhaustion clouded Brian's brain.  His concentration drifted away.  He checked his altimeter that indicated their descent through 5,000 feet, well below the 10,000 foot altitude where oxygen was required.  As he had done many times before, Brian unsnapped his oxygen mask from the left side of his leather helmet.

The feel of real air temporarily refreshed him. The faint smell of gasoline, oil and even a little ethylene glycol coolant made the air more real. The moisture soothed his throat.

They landed at RAF Middle Wallop with what had to be only a few gallons of fuel. The crews would know shortly when they filled up the tanks. As Brian taxied his 'PR-F' Spitfire to a stop in his parking spot and switched off the engine, Gordon, Toldson and Jenkins scrambled around the fighter. Brian moved his goggles to his forehead, lay his head back against the rest and closed his eyes. Every muscle in his body relaxed, as he took a deep breath and exhaled slowly. Sleep coaxed and seduced Brian into letting go. It felt so good, so comforting, so warm.

"Are you all right, Mister Drummond?" asked Bernie Gordon, pulling Brian back to reality.

"Just tired, Bernie," Brian answered, keeping his eyes shut. "Real tired."

"We've had a rather busy time of it as well. Just received a few presents from our German friends. Nothing serious," he said, as he performed his turnaround checks on the aircraft. "And, we've had a few visitors all day. Seems we have fighters scattered all over the South like you chaps did earlier over at Tangmere." Brian barely nodded his head as sleep began to gather him up again. "We shall have her ready to go in a few more licks, and then we will leave you alone to take a quick nap."

Brian did not care about anything else as his fatigue claimed his consciousness.

Something dragged him through the thick, murky substance of bone numbing fatigue. He did not want to make the journey. He wanted to stay in the warm comfort of his unconsciousness, but something persisted with a clarity of purpose.

"Sir." He heard the word in the distance. "Sir." There was it again, only more distinct this time. Then, in rapid succession, he felt a hand shaking him, the characteristic tones of several Merlin engines in the process of coming to life, and light and color returning to his brain.

"Sir," said Bernie Gordon. Brian finally opened his eyes, only to see the worried expression of Leading Aircraftman Bernard Gordon, his crew chief. "Sir, you have another mission. You're full up on fuel, ammunition and oxygen. The bird is in good shape."

"Thanks, Bernie," Brian gurgled, as he cleared his throat, pulled himself back upright in his seat and checked his straps, connections and switches.

Brian moved calmly and quickly through his start procedures as his mind wondered how long he had been asleep. Everything around him remained as he had left it, except for the aircraft beginning to move for takeoff. Beamish and Kradilcek waited for Brian to catch up, and then Green Section taxied to the takeoff point together.

The squadron headed toward the southeast. They were passed from Middle Wallop Sector Control to Tangmere Sector Control for assistance to No.11 Group. As they reached the coastline, they turned east to parallel the beaches. The squadron leveled off at 20,000 feet before they were passed to Kenley Sector Control.

Kenley provided a quick status report. Numerous late afternoon raids were trying to escape back to occupied France. No.609 Squadron gave chase. The sky was laced with dissipating streams of black and white. Clouds of dust and smoke dotted the landscape below. It did not take long to find the retreating intruders.

"Tally ho," broadcast 'Spike' Darling. "We have birds and bees at various altitudes headed south."

"Engage at will."

"Here we go, lads. Run 'em up. We'll give Gerry a run for his money."

Pilot Officer Brian Drummond pushed his throttle through the emergency gate as the other pilots did the same. Darling rolled the squadron into a chase attack position. They could see the fighters weaving above and behind the bomber formations. The bombers without their heavy bomb loads proved fast enough. The coast of France appeared through the clouds ahead of them. Brian knew the call to return would soon come. After Dunkirk, the prohibition of flight over enemy territory had been reinstated. In fact, it was quite uncommon for them to be out so far over the Channel.

As they reported their inability to catch the retreating Germans, Kenley Sector Control indicated several stragglers were scattered about trying to make it back to their bases. They descended into the haze as the late afternoon sunlight reflected back to them making visibility even more difficult. Several vectors were received as they tried to find their targets. Frustration mixed with fatigue produced a vile brew that eroded any remaining patience. They trolled through the murky air as long as their fuel permitted.

Darling and Kenley Sector Control eventually agreed, and No.609 Squadron was allowed to return to RAF Middle Wallop. In the oddest manner, Darling began to chatter away over the radio trying to engage his pilots in enough humorous recollection over the day's events. They appeared to be the only squadron still in the air as far as they could tell, and it was Darling's way of helping his pilots and probably himself to stay awake until they landed safely at their home base. The pilots resisted at first, and then became caught up in the unusual opportunity. They watched one another for telltale signs of any nodding off – a wing falling off, a drifting departure from the loose formation, erratic movements or slurred speech. Even the controllers on the ground participated. The purpose of the strange chatter did not need to be explained.

The normal return of a completed mission did not take long as everyone understood the fatigue. Everyone was exhausted. Even Corporal Warren had difficulty keeping her eyes open. The pilots began dropping like flies. A few never made it to the last serving of the evening meal. Some fell asleep at the table with plates of uneaten food. Brian ate a few bites of his food that he could not recognize before he trudged up the stairs. He started to take off his sweat soaked tunic and shirt. Only half his shirt buttons were undone before he collapsed into his bed.

———

*Monday, 19.August.1940*
*RAF Middle Wallop*
*Middle Wallop, Hampshire, England*
*07:00 hours*

**B**rian Drummond and Jonathan Kensington sank into the lawn chairs outside their dispersal tent. They dragged through each step as if they silently suffered pain with every movement.

"Did somebody do something to time?" asked Jonathan with his eyes closed.

"Yeah," answered Brian. "I think God took away the nighttime to see how tired we can get and still fly."

"I have never been so tired in all my life."

Several of the others mumbled something. Brian glanced around him. All the pilots, whether sitting in chairs or lying on the still damp grass, had their eyes closed. Only Darling and Beamish had fresh uniforms. Brian, like probably several others, had not even bothered to change clothes. The telltale white borders of yesterday's sweat stains marked the extent of their fatigue.

"How much longer can we keep this up?" asked 'Boxer' Stockard.

"As long as we have to," Darling said with a twinge of defiance.

"They are going to grind us down," added Kormer Mansek, "like they did in France."

"The fight's not over, yet," Stockard said.

"You tell 'em," mumbled 'Jackstay' Beamish.

In a few minutes, only the chirping of the birds could be heard as one by one the pilots found sanctuary in sleep. Their respite was interrupted by the launch commotion of the No.238 Squadron 'VK' Hurricanes on a mission of unknown, but probable, purpose. The variety of missions had narrowed to two – redeployment or combat. There was far more of the latter than the former.

Brian was asleep when the telephone rang, and they were brought to Standby status. Before all the pilots made it to their cockpits, the launch command came. They were in the air in a few minutes.

No.609 Squadron joined a confused air battle over Southampton in support of No.11 Group and the Tangmere Sector squadrons. As they leapt into the fighter portion of the battle above the bombers and the determined Hurricanes, the signs of concentrated bombing of Southampton's piers, docks and port facilities provided the backdrop to focus their purpose. Black smoke covered most of the city. As they positioned for their entry into the fight, Brian glanced down through the clouds to see the distinctive saw-tooth roof of the main Spitfire factory across the River Itchen from Southampton. There were no signs of any bombing around the Supermarine plant. Brian shook his head with the miraculous good fortune that the Germans did not know what vital activity went on below the large jagged roof.

The fight evaporated as they joined. The bombers turned and headed south while the fighters covered their withdrawal. Brian, as well as most of the others, managed to find several fleeting targets to fire at without consequence.

The clouds thickened substantially by the time they recovered back at RAF Middle Wallop. They passed through several layers of clouds and made a couple of course adjustments once they were underneath the bottom cloud deck.

The darkening clouds did not alter the intensity the crews displayed in turning around the fighters. The pilots returned to their struggle to shake the tentacles of fatigue gripping each of them.

Flying Officer Royster returned from the Operations building. "Weather forecasters say this is the beginning of several days foul weather," he said loudly without an attention cue to see if anyone was listening. Several pilots including Brian opened their eyes.

"Hallelujah."

"Praise the Lord."

"Our prayers have been answered."

"The main part of the storm won't be here until late tomorrow," Royster said somewhat surprised by the reaction.

"Not to worry, James," said Flight Lieutenant 'Sparky' Morrow with his eyes still shut, "this lot of bloody, stone cold, killers actually enjoys the tension of mortal combat."

"Here, here."

"We are actually quite disappointed," Morrow continued.  "We live to fly, and we fly to live."

"Well said, old boy," added Flight Lieutenant 'Waggle' Davies, the freshest of the pilots, having just returned early from his convalescence.

"Not so fast, you lot," said Squadron Leader Darling, as he joined them inside the tent.  "Let's mount up."

"But, the tellie didn't ring," protested 'Boxer' Stockard.

"Right.  I called them."

"Oh, Skipper," responded Stockard.  "Now, we have to go looking for trouble."

"Is our mission in life," interjected 'Crazy' Kradilcek.

"As the man, says," Darling said, "this is our job.  Now, plug the yappers, gather your kit and mount up."

The pilots did as they were ordered.  Brian scanned the growing ominous-ness of the darkening sky.  This was not going to be an easy flight with 12 fighters groping through turbulent clouds trying to find an illusive enemy.  What were the bombers going to see anyway?  While they were aware of the rumors about some radio beam aided bombing system the Germans had, none of them had ever heard of the Germans dropping bombs in bad weather where they could not see the target.  Maybe this was a change in their tactics?  Maybe there would be protection in the winter weather?

As Brian predicted, the flight was a struggle just to maintain position on the others.  They entered the lower cloud layer at about 2,000 feet and did not breakout until nearly 11,000 feet.  There was at least one more layer several thousand feet above them.  If the next layer of clouds had any depth, they would be immersed when they arrived at their assigned altitude.  The flight entered the clouds again as the pilots groped for something more.  Sector Control's radio calls took on an unusual air of frustration as the squadron of fighters leveled off at 15,000 feet in the clouds.

They tried above and below the clouds turning in various directions.  Brian could not pick up even a whiff of scent although Sector kept telling them they were over the top of the bandits.  An evil potion of frustration, fatigue, resentment and fear began to corrode the concentration of the No.609 Squadron pilots. Radio calls became sharper. The formation began to lose its discipline. They hunted for the Germans for more than an hour as the clouds thickened and darkened. Every time they entered the clouds they all had to be asking themselves the same questions Brian was asking himself. Were they going to find the Germans in one of these cloud layers? Would they find them through collision? Where were they? The longer they flew the more disoriented Brian became.

Finally, mercifully, Sector Control called off the hunt and gave them vectors to the south. They would get below the clouds over the Channel where there were no hills and buildings, and then they would work their way back to RAF Middle Wallop at low level.

Normally, Brian would have enjoyed the rush of high-speed, low level, flight across the countryside. Visibility was dropping rapidly. The navigation task proved to be far more difficult than any of them wanted. Fuel levels thinned to the point of concern as they made numerous course corrections trying to find their airfield. Brian knew the concern when Squadron Leader Darling asked if any of them saw a sufficient field for them to put down in before their fuel was exhausted. They received no help from Sector Control. Just as Darling decided to break up the flight to allow some aircraft to get on the ground safely, Flying Sergeant 'Fog' Johnson found the airfield. One by one, they landed.

A light drizzle forced them into the tent with its two hanging bare bulbs for light. The intelligence folks were nowhere to be found. Darling had each of the pilots fill out a debriefing form themselves. No words were spoken for the longest time.

"I don't care to do that again," said Flight Lieutenant 'Sparky' Morrow, the second most senior officer in the squadron.

"You are bloody well right there, old boy."

Squadron Leader Darling stood at the open, flapped entrance to the tent gazing into the distance through the drizzle. Brian could only surmise what might be running through his mind. They all knew weather was one of the risks that any pilot took, but compared to combat, it always seemed so trivial and unnecessary. They wanted to save their luck for important things.

Despite the deteriorating weather, the squadron remained on alert throughout the remainder of the day. The pilots took advantage of the respite. Most slept in the chairs. The bright spot of the day was the arrival of 'Spike' Darling's wife with a tray of sliced chicken sandwiches and hot tea. They missed the noon meal at the Officer's Mess. Emily

Darling added a delightful light to the dreary tent.  Everyone became more animated and upbeat when she was there, and returned to their solemn state when she left.

—

*Monday, 19.August.1940*
*The Admiralty*
*Whitehall, London, England*
*14:00 hours*

"**M**ister Trevor Andersen is here to see you, Admiral," announced Sir Geoffrey Pike's clerk.

"Please show him in," 'Jumper' replied.

The 36-year-old, Trevor Thomas Andersen had been working for Naval Intelligence for ten years as a field agent.  He spoke and wrote fluent German and Polish.  Trevor was also known as Agent 'Diamond' or just 'Diamond,' and often used the alias Robert Henry Stone Johnston for his clandestine activities in Germany and Poland.  His light brown, wavy hair, ice blue eyes, and chiseled features gave him a rather model Aryan appearance, further enhancing his adaptability to operations in and around Germany.  Vice Admiral Sir Geoffrey 'Jumper' Pike had personally recruited Andersen into Naval Intelligence, shortly after Trevor's graduation from Cambridge University with a degree in European history.

"Great to see you again, Admiral."

"The honor is mine, Trevor.  Close the door and have a seat," Sir Geoffrey said, gesturing to the chairs around a small table to the right of his desk.  Andersen did as he was asked.  The two men shook hands and sat across the table from each other.

"I don't think I have seen you since the evacuation in June, so please allow me a belated congratulations and well done for your assistance to the Earl of Suffolk for the extraction of the people and material from France."

"Thank you, sir.  That was a pretty hairy endeavor with the Germans breathing down our necks outside Paris."

"Most folks will never know how important that operation was to our war effort."

"Glad to be of service, sir."

"That said, I wanted to see you directly to tell you the good news." Andersen nodded his head and waited for Pike to continue.  "Several weeks ago, we finally made contact with 'Blocker.'"  Actually, 'Blocker' was Colonel Stanislaus Pordonski of the Polish secret police, the man who led his team of operatives with Andersen to stage an ambush and capture a German, three-wheel, Enigma cipher machine in 1939, before the war began.

"Excellent. Where is he?"

"Below deck," Pike answered in his nautical jargon.

"I'll be damn. How is he?"

"You will see for yourself as soon as we are done here. He is nearly finished with his debriefing, so this is an opportune moment in several ways. I wanted to talk to you about the future. We have not discussed this and I imagine you are aware of the War Cabinet's recent decision to form the Special Operations Executive."

"Yes, I am."

"After some protracted discussions with Hugh Dalton, we came to an end point. In short, he wants your skills and experience in SOE, and I do not want to lose you."

Churchill and the War Cabinet had appointed the Right Honorable Edward Hugh John Neale 'Hugh' Dalton, Member of Parliament for Bishop Auckland, to serve as Minister of Economic Warfare and to be chief of the Special Operations Executive (SOE), newly created just in the previous month. Dalton was also another Labour Party member to attain ministerial rank in Churchill's coaltion government.

"Thank you for your confidence in me, Sir Geoffrey."

"Nonsense, Trevor. I would be a fool to let you go. That said, Hugh and I agreed to leave the decision to you. To be frank and candid, I think the SOE would offer you more of what you are best at doing and I suspect you most enjoy doing."

"I suspected that was the subject you wished to discuss. I have considered the question. At least as I understand things, I believe you are correct. I think the SOE is more to my interests. However, my loyalty remains with you, sir. You saw performance within me that I did not see myself. You have been an exceptional mentor. I wish to stay with you."

Admiral Pike smiled. He cherished such loyalty. However, Sir Geoffrey was also a realist and pragmatist. "Thank you for your loyalty, Trevor. I am approaching old age. I will not be in this job forever. I serve at the pleasure of the King, and I could be dispatched tomorrow. As you give me your loyalty, so I shall give you mine in full."

"Perhaps, I can serve you both."

Sir Geoffrey chuckled softly. "You cannot serve two masters, Trevor. That aside, I shall discuss a modest refinement of your assignment to SOE, to allow collateral taskings from Naval Intelligence.  I truly believe you will be a masterful operative with whom we will all be proud."

"If that is your wish, sir, then I am so commanded."

"Very well, then, we have an agreement.  Now, with that done, let us discuss 'Blocker.'  The enemy occupies his country.  Do you think Pordonski would be amenable to working with you at SOE?"

"I think he would appreciate the opportunity to take the fight to the enemy.  However, that decision belongs to Mr. Dalton."

"Yes, quite so, but I am asking if you would endorse his service in SOE until Poland is liberated?"

"Absolutely, sir.  I have witnessed his work more than a few times.  He is a natural."

"Excellent . . . as I suspected.  I will recommend that to Hugh.  First, when you talk with Pordonski, please ask him for his preference.  If he is agreeable, then you can discuss it with Minister Dalton, when you interview with him."

"Yes, sir . . . as you command."

"Now, let us get you reunited with your friend."

———

*Monday, 19.August.1940*
*The Admiralty*
*Whitehall, London, England*
*15:15 hours*

Trevor knew the way to the basement interrogation rooms.  There were only four such rooms, so the process of elimination did not take long. He found the supervisor of the interrogation team.  They were complete and only socializing,, as Admiral Pike had notified him to entertain Pordonski until Andersen arrived.  The supervisor led Trevor through the security doors.

"Trevor," boomed Pordonski, as he stood, stepped toward his friend and extended his right hand.

As they shook hands, Andersen said, "Great to see you again and that you are safe."

"I am, and thanks must go to His Majesty's Government."

The two interrogators excused themselves and suggested they use the floor conference room, as the door could be opened from the inside.  They moved, found some tea and cookies, and took adjacent seats on a corner of the long rectangular table.

"How the hell did you get out?" asked Andersen.

"It is a very long story, my friend."

"We have plenty of time. We can continue through supper and drinks."

Pordonski smiled. He began in Polish. "As the Germans crossed our sovereign border a year ago, I became rather busy, as you can imagine. To be frank and blunt, we were no match for the Germans . . . their aircraft, their tanks, their overwhelming numbers. We simply could not stop them. We fought with the infantry into the center of Warsaw. It was not pleasant. Then . . ." he paused, closed his eyes, and lowered his head, as if he was recalling a most depressing moment. "I remember the moment, precisely . . . ." Pordonski swallowed hard. "It was 08:07, Wednesday, the 27th of September – the day Warsaw fell and my government surrendered to the invaders. I turned around and there was no one there. They all evaporated, leaving me standing there with my cock in hand, so to speak – a very disappointing moment – a revelation of sorts, and not the good kind. Fortunately, the Germans don't fight much at night. It took me several months just to make it to the Ukrainian border, only to face the damn Red Army. I was wounded twice in a skirmish with the Reds. Fortunately, I still have friends among the Ukrainians, and they are not particularly fond of the Russians. They got me medical treatment, hid me, protected me, and then helped me move. Romania was more problematic than I expected. It seems they may be more closely aligned with the Germans than any of us imagined. Anyway, the winter and spring passed with small steps from house to house, until I finally crossed the Bosporus last month and sought refuse in the British consulate in Istanbul. I invoked Admiral Pike's name. From there, MI6 and Naval Intelligence had me. They had to be sure who I was and the sincerity of my condition. I do not blame them for their caution, but I did not feel it reasonable to connect with you through unknown people."

"The Naval Intelligence people would have been trustworthy, and probably the MI6 lads as well. Regardless, thank you for your caution, but I am sorry I was not there to help you."

"There are rules, my friend."

"Yes, there are . . . but not among friends."

"They moved me through Cairo, Malta and Gibraltar, before I arrived in Liverpool and finally London."

"I am truly sorry you had to go through all that, Stan. But, you are here now."

Pordonski laughed a deep, hearty laugh. "Indeed, I am," he said, holding his broad grin and lightness to his tone. "I am alive, and I live to fight another day."

"Yes, and that is also why we need to talk about the future."

"How so?"

"The Polish government in exile has been established here in London. As an officer in the Polish secret service, so I suppose you are still a member of the government. However, there may be other opportunities."

"What have you?"

"Last month, His Majesty's Government created a new organization called the Special Operations Executive, under the direct supervision of Minister of Economic Warfare Hugh Dalton. The mission of the organization is unconventional warfare on the Continent to disrupt German military, political, economic and social operations and activities."

"Sounds very interesting."

"I thought so as well. Admiral Pike has transferred me to the SOE earlier this afternoon. He thought you might be interested."

"Most assuredly," Pordonski answered in perfect English.

"Excellent," Trevor said in English as well, and then he switched to German. "Are you ready to be German . . . to kill Germans?"

"As German as you, my lethal comrade," Pordonski replied in German.

"One last question to be clear. You prefer to join SOE rather than work for the Polish government in exile?"

"Yes. And, to be equally clear . . . the best way for me to help my country is to help the British defeat the Germans."

Andersen switched back to Polish. "Excellent. I shall inform Admiral Pike. I expect to get an invitation to interview with Minister Dalton. I will try to get this settled quickly." Pordonski nodded. Andersen switched back to English, again. "Now that our business is done, let us go find a nice supper and some entertainment. We have much to remember from the good times."

"I am most agreeable to that plan."

The two men departed to begin their evening and extend their friendship.

—

*Monday, 19.August.1940*
*No.10 Downing Street*
*Whitehall, London, England*
*17:00 hours*

**P**rime Minister Churchill had been feverishly working through the daily correspondence with John Martin and two stenographers, actually six stenographers in relay, two at a time. The knock on the door preceded the entry of his Assistant Private Secretary 'Jock' Colville.

"The Foreign Minister has arrived, sir."

"Very well, please show him in." Winston looked to Martin. "If you will excuse me, John, I am afraid this is going to be a private conversation."

"By all means, Prime Minister," he said, and then gestured to the stenographers to leave.

As Martin and the others left his office, Lord Halifax entered without waiting to be introduced. Colville looked in with a shocked expression on his face for the impropriety of the Foreign Minister's impatience. Churchill shook his head . . . not to worry about it. Churchill motioned to chairs to the right of his desk, as he rose to join Lord Halifax.

The Foreign Minister extracted several sheets of paper from his carrying case. "Lord Lothian received a handwritten note from President Roosevelt to you, just after midnight this morning, while he was visiting the President at his Hyde Park estate this weekend. After discussion with Lothian, we agreed to have it transcribed in New York City by Stephenson's communications people for immediate transmittal to us. So, if you will forgive me . . . ," he said, handing Roosevelt's cover letter to Prime Minister Churchill.

---

*Hyde Park, N.Y.*
*August 19, 1940*
*My dear Churchill:-*
*    I think this will interest you. It was over two weeks on its way from Berlin and coming from an American, long a resident in Germany, it has especial value. The writer was, I think, inclined to be pro-Nazi up to the time of the Munich Conference.*
*    We are getting excellent reports of the fine job your Air Force has done the past week.*
*    As ever yours, FDR*

---

"The letter to which the President refers was sent to him personally a few weeks ago by the American *Charge d'Affaires* in Berlin, Alexander Comstock Kirk, whom I have known personally from his years of service in the diplomatic corps. He has been in Berlin since May of last year and is now the senior State Department diplomat in Germany. You may recall, the Americans recalled their ambassador after *Kristallnacht*. He is a good man, very observant and astute. Needless to say, a *charge* communicating privately with the President of the United States is rather unusual; however, I think you will understand why he did so." Lord Halifax handed the Kirk letter to the Prime Minister.

---

Berlin, July 29, 1940.
My dear Mr. President:

I fully realize that you are not to be importuned with lengthy or superfluous messages, but there is a matter much on my mind and one, which I dare not treat by telegraphic report.

It is perhaps anomalous that at a time when the entire world seems to be concentrated on the mechanics of war the main emphasis in rumor and report, at least insofar as Berlin is concerned, has lately been on the subject of peace between England and Germany.  It is futile to argue about the possible origin of these rumors for it is as easy to find support for the opinion that they are inspired as it is to believe that they are the manifestation of a natural inclination on the part of war-tired peoples.  It is useless also to test the sincerity of these expressions, for again it is not only impossible to place the responsibility for their utterance on any particular source, but also to strike an accurate balance of the relative advantages from the German standpoint of an immediate peace with England. The fact remains that talk is insistent on this subject and within the last few days three private individuals, an American, a Hollander and a Swede, have brought me stories of peace efforts allegedly sponsored by high Nazi officials, but decidedly not by Hitler himself — stories which I could not report in detail owing to the certain danger involved to the individuals themselves.

The purpose of this letter, however, is not to evaluate these rumored activities in relation to Nazi aims or policies.  My purpose is solely to register my profound conviction that any concession on the part of the British Government now would destroy forever the chance of eradicating the forces, which are threatening our own civilization.  In saying this I am not thinking of England itself for,

if I did, I might hesitate at the thought of
the terrific devastation, which may well be in
prospect and of which the beginning is now
being set from week to week.  Hitler has always
tried to attain his aims at the lowest possible
cost to himself and has so far triumphed in
that policy, but there has never been any doubt
that in the last analysis he would and he must
pay any price to attain his ends, if they can
not be otherwise achieved.  There is every
reason to believe that he wants to finish this
particular phase of the war, whether for the
purpose of turning his war machine in other
directions or of indulging in expressing his
colossal ego through rebuilding in his own way
on the ruins of what he will have destroyed.
A short war requires, in the first place, the
speedy subjugation of England either through
the peace he might be ready to negotiate now or
through the same or worse methods than those
which prevailed in France, and, in the second
place, the certainty that with England silenced
the forces of democracy would be annihilated.
As regards the first requirement, the greatest
triumph for Hitler would be a humiliating and
unjust peace wrenched from England without a
fight, for the hope is justified that even the
conquest of the British Isles need not end
resistance. As for the second requirement the
greatest factor is our own country and, I firmly
believe, the controlling factor.  Hitler has
never lost sight of the United States although
there are times when he believes that he can
disregard our part on the ground of the time
element involved.  Within the last few weeks,
however, I feel that his anxiety on our account
has redoubled.  He sees what we are doing and
that we shall not stop, and he must know that
we constitute a problem, which even he cannot
solve.  It is natural to say that there is
a limit to a one-man show and that in time
Hitler must eradicate himself.  That may be

true but the rate of his progress reduces the
saving grace of the time factor and the force,
which he has developed, can be stopped only by
force.  The first stand in this struggle is for
England to hold out against any peace efforts,
and the second is for the British to exert all
their material and moral equipment to resist
an attack and continue the fight. The greatest
part, however, I believe is ours.  We must
encourage in every way those who are in the
first line of battle.  We must prove our purpose
in fact and example, and we must prove it
speedily and unflaggingly.  We must hold to the
principles, which we know are right although
their implementation may require newer methods
and at the right moment we must be ready to say
the word and do the deed that will save from
destruction all that we know makes life a good
and noble thing.  It is the conviction of the
significance of our part in the struggle, both
for our own salvation as well as for the benefit
of the world, that induces me to write and I
hope that you will consider what I have said in
that light.

I cannot close without expressing to
you my gratitude for the consideration which
you showed me during my stay in Washington,
and I need not say how gratified I am that
the continuity of your policies, of which we
ourselves were always assured, has now been
brought to the conviction of the governments
and peoples abroad.

Very faithfully yours,  *Alexander Kirk*

---

Churchill looked to Halifax but did not speak.

"What makes the Kirk letter far more poignant, Winston, rests upon the fact that Alex believed as we did that the Germans had a right to redress the injustices of the Treaty of Versailles.  He has quite apparently changed his mind regarding the intentions of the Nazis."

"I am gobsmacked, to be frank," Churchill said.  "Quite a contrast to Joe Kennedy, isn't it?"

"Yes, quite so."

"The President is telling us indirectly he does not trust Kennedy and certainly does not agree with Kennedy's defeatist opinions. In the light of the publicly announced agreement between Canada and the United States for hemispheric defense, and the President's decision to lend us his surplus destroyers, this letter tells me, we have the clearest statement to date of the President's intentions to help us. He certainly had no obligation to share the Kirk letter with us, but he did so with purpose. He is encouraging us and telling us he is not alone in that endeavor."

"That would be my assessment as well, Winston."

"Ed," Churchill said, using the contraction of the viscount's given name, "if you would be so kind, please send a simple reply to the President with our gratitude for his generosity. Also, message Philip and Bill thanking them for their extra effort with this communication."

"I will certainly do so on your behalf. If I may say so, Winston, it feels like we are turning the corner."

"My sense as well, although we must beat back the threat of this damnable invasion before I think we can safely say so. I am certainly feeling better about our prospects than I was just a few weeks ago."

"Is there anything else we need to discuss?"

"No, Ed, not at the moment, I should think."

"Very well, then. By your leave . . ."

"Thank you, Ed."

The Foreign Minister departed as swiftly as he had appeared. Churchill took a moment to stare out the window of his office at the magnificent residential garden behind No.10. He wondered whether the nexus of British defense and American industrial support would come soon enough to preserve this garden and his country. After several minutes, he called Colville to ask Martin and the stenographers to return. They had work to do.

———

*Monday, 19.August.1940*
*RAF Middle Wallop*
*Middle Wallop, Hampshire, England*
*17:45 hours*

The drizzle had turned to rain by late afternoon when they were finally released from alert. This felt like a night for camaraderie. They went directly to the Officer's Mess bar to drink a few beers or other spirits before the evening meal. Squadron Leader and Mrs. Darling joined them. Laughter returned to the bar, as they reverted to their less mortal pasttime and the experiences that brought them through history.

The appropriate break came as evening meal neared. Each of them retired to the residence, showered and donned a fresh set of clothing. The moment they showed up for dinner turned back the clock six months, when events seemed to be less serious. Humor and enjoyment of life blotted out the fear, horror and ugliness of the real war. Even Emily Darling could laugh at the momentarily distant threat to her husband.

The evening's impromptu celebration of life continued in the bar. No one displayed any interest in braving the rain. The venue did not dampen the spirit. Even 'Slim' Koenig and 'Crazy' Kradilcek were laughing again. The stories of close calls, mistakes past, unique experiences and the mishaps of others filled the spaces between the drinks and laughter.

They heard the Blenheim night fighters of No. 604 Squadron take to the air. The sound invariably brought curiosity and wonderment how those pilots launched into the black of night, trying to find invisible German bombers on their nightly journeys of destruction? Word came quickly they would intercept an inbound raid, this night headed for what appeared to be the Midlands or perhaps the dockyards of Liverpool. If past experience was any judge, they would remain aloft to engage the bombers on their return to France.

Emily and Horatio Darling were the first to retire. The others continued into the depth of night mostly to complete the release of their inner tension, but also as an unspoken act of defiance that would hold back tomorrow. Fatigue eventually claimed them all.

———

# Chapter 6

There is a road from the eye to the heart
that does not go through the intellect.

-- G.K. Chesterton

*Tuesday, 20.August.1940*
*RAF Middle Wallop*
*Middle Wallop, Hampshire, England*
*09:15 hours*

### Week 7

The rain continued in a steady, soaking fall that turned the grass of the takeoff and landing area into a soft, squishy field incapable of supporting a fully loaded, 6,700 pound Spitfire fighter.  Although they had no means to get airborne and the low clouds, constant rain and occasional rumble of thunder told them not to fly, they waited as they always waited when they were not in the air.

Brian's thoughts in between short conversations with the other pilots turned to Charlotte Palmer.  He felt an overwhelming need for redemption and forgiveness.  At the first opportunity, he decided he would make another visit to Standing Oak Farm to offer his apology and beg for her absolution.  The connection with her proved to be more important to Brian than any of the more basic urges.  Although Malcolm Bainbridge had pulled him from the wreckage of the Stearman two years earlier and the sea rescue crew extracted him from the cold, numbing waters of the English Channel two months earlier, Brian never realized, or maybe recognized, any proximity of death.  Charlotte Palmer saved his life in a different, more personal way.  She reflected his mortality like an angel summoned to alter the course of his destiny.

A young aircraftman arrived with the morning mail.  It was a bountiful day for most of them.  Even 'Curly' Mansek and 'Crazy' Kradilcek received letters from relatives who fled the Nazi invasion to Sweden, Switzerland and England.  Brian was no exception.  He received two letters, one from his mother and one from Mary Spencer.  He held Mary's letter that smelled of her perfume.  The pink stationery and delicate, careful handwriting told him he needed more privacy.  Her letter could not be good, no matter what.  He slipped Mary's letter into the inside pocket of his uniform tunic, and then opened his mother's letter.

*July 14, 1940*

*Dear Son,*

*Thank you for your letter of June 20th. We always seem to rejoice when we hear from you. We miss you so much, and can only pray that you are well.*

*The newspaper accounts of what they are calling the Battle of Britain are almost impossible to read. Your experience during the Battle of France and this horrible war makes me shake with fear for your safety. I know I promised not to bother you about this adventure you have chosen, but Brian, you are our only son, and this war is so brutal. I need to change the subject.*

*Dad's business is doing fairly well. He is working very hard. I am helping in the shop when I can, but he could really use your help. I am working with several other women on various community projects. We are trying to make Wichita a more beautiful place.*

*We saw Becky the other day. She is home from Lawrence. She looks good, seems to be very happy and doing rather well at the University of Kansas. I wish things had worked between the two of you. I really like her. She is a very pleasant, caring young lady. But, what will be, will be.*

*Everything else is clicking along real nice. It is just I worry about you all the time wondering if something has happened to you over there.*

*Oh, Gertrude Bainbridge wanted me to tell you hello, and she sends her best wishes for your safe return to us. She tries very hard, but she is not the same lady she was before her husband died. I don't want to be like that, Brian. Don't take any unnecessary chances. I remember what the last war was like to our young men, and I don't want that to happen to you. Please promise me you will take care of yourself.*

*We miss you very much, Brian, and we pray for your safety. Please write when you can. I know you must be very busy, but it is so nice reading your words. We love you very much and miss you even more. Please take care, Brian.*

*Love,*

*Mom & Dad*

---

Brian refolded his mother's letter, returned it to the envelope and slid it into the pocket along with Mary Spencer's letter. He sat in his chair and stared out the tent entrance at nothing in specific.

Jonathan Kensington moved over to the chair next to Brian. "What is it this time?" he asked in a near whisper.

Brian looked into the eyes of his best friend, but did not speak for several moments. "A letter from my Mom."

"Bad news?"

"Naw," Brian responded, smiling toward the rain outside. "I just miss me Mum, as we say around here."

Jonathan chuckled. "Well, then, good to know 'Hunter' is normal."

"Yeah, thanks."

They both laughed and settled into their thoughts. Brian's focus narrowed to the other letter inside his tunic. His curiosity began to add anticipation for what might be in the letter. He also thought of Charlotte Palmer, Rosemary Kensington, and as was usually the case, Anne Booth. While the intrigue surrounding Anne Booth added an unsavory dimension to his life, the complications that Mary Spencer represented defied comprehension. Somehow, he had to find a more stable relationship with her.

Brian noticed the rain had stopped, at least for a short time. He rose from his chair and walked outside. The damp, heavy air possessed a refreshing coolness that invigorated Brian. The wet grass and soft ground offered a familiar earthy, farm aroma to the air around him. He walked past the right side of the tent to the short hedgerow that marked the southern boundary of the aerodrome. He looked to the dark sky above, and then considered what might be in Mary's letter, looked around to see if anyone would join him, and retrieved her letter.

---

*17th August 1940*

*Dearest Brian,*

*I truly hope this letter finds you safe and in the best of health. I miss you more than I surmise you will ever imagine. I also would like you to know that I recognize the difficult predicament I have placed you in, and for that I must apologize. While I do love my husband, I find myself attracted to you in a way I have not felt in many years. I have no intention of hurting you or John, but I find myself thinking of you constantly.*

*My confessions were not the intent of this letter. As much as I know it troubles you, I must see you as soon as possible. I have some very important news I simply must share with you, and I will only*

*discuss it if I can see your eyes.  It is fabulous news, so not to worry.*
*Please call me as soon as you can, so we can arrange a meeting place*
*and time.*
*With Deepest Love,*

*Mary*

---

Brian held the letter as he looked at streaks of rain from the clouds to the southwest.  His thoughts could not imagine what Mary's purpose and news might be.  What did she want from him?  What could he possibly give her that John Spencer could not?  Why couldn't she leave him alone?  She said it was good news, so it could not be an illness, tragedy or injury.  Should he try to see her or not?

"What now?" asked Jonathan, startling Brian.  "Rather jumpy, aren't we?  Must be a good one."

Brian folded the letter, slipped it into the envelope, and then returned it to his inside pocket.  "It was from Mary Spencer."

"Brian, I thought you stopped that."

"I did, but she says she needs to talk to me eye-to-eye."

"Are you going to do it?"

"She says it's important."

"They all do, Brian."

He looked off at the approaching sheet of rain.  Jonathan stood beside him watching the same clouds.  What could he say to Jonathan that could justify seeing Mary Spencer?  Brian felt the tug of the whirlpool only now ever so slightly pulling him to its center.

"Ay, you two," called Flight Lieutenant 'Jackstay' Beamish.  "Skipper wants to talk to us.  Shake a leg."

As the two friends entered the tent with the others, Squadron Leader Darling began, "Group wants each pilot to read a special instruction from One One Group.  I must remind you this letter is designated as secret.  I shall pass it around," he said, as he handed the single piece of paper to Flying Sergeant 'Fog' Johnson.  "Air Vice-Marshal Brand asked me to convey a few amplifying words after each of you reads the memorandum."

Brian watched for any expression on the faces of the other pilots as they read the message.  There were no clues as to whether the content was good or bad.  The fact that each pilot was asked to read a secret instruction from a neighboring group most probably meant the information had to be significant.  Brian was next to last to read it.

HEADQUARTERS, No. 11 GROUP

FIGHTER COMMAND

ROYAL AIR FORCE,

UXBRIDGE, MIDDLESEX

Telephone Nos.  UXBRIDGE 2894 (4 lines)

UXBRIDGE 2896 (2 lines

Telegraphic Address: "AIRGPLON UXBRIDGE."

Reference: -- FC11/P.11048

# SECRET

19th August, 1940.

GROUP INSTRUCTION No. 4

This instruction is issued with immediate effect and shall remain so until further notice.

2.   Controllers shall endeavour to dispatch fighter sorties so as to engage enemy aircraft over land.  Rescue forces have been deployed to provide prompt assistance to downed crews.

3.   Controllers shall make every effort to avoid sending fighters out to sea.  Pilots shall break off engagements with enemy aircraft over the sea and especially with enemy aircraft withdrawing over the sea.

4.   Controllers shall dispatch two fighters to engage single enemy reconnaissance aircraft.

5.   Controllers shall dispatch the minimum number of squadrons against mass raids.

6.   No.12 Group will be asked to protect No.11 Group aerodromes while fighters are airborne.

7.   If heavy attacks head inland, controllers shall place a squadron or training unit over aerodromes especially under low clouds.

8.   No.303 (Polish) can provide two sections for patrols over inland aerodromes.

```
            9.  No.1 Canadian Squadron (RCAF) can be
    used for daylight operations.
                    As ordered by,
                      Keith Park
                    Air Vice-Marshal,
              Air Officer Commanding-in-Chief
      No. 11 Group, Fighter Command, Royal Air Force
```
**SECRET**

Brian waited for Pilot Officer 'Boxer' Stockard to finish his reading. He did not understand the significance of several of the points made in Air Vice-Marshal Park's instruction and hoped Squadron Leader Darling would explain the importance.

"As undoubtedly you can appreciate, One One Group has taken some heavy attacks since the German's so called Eagle Day. The intentions of the enemy are now clear. They are likely to make every possible effort to obliterate the southeast aerodromes to gain local air superiority for an expected invasion attempt."

"How soon?" asked 'Fog' Johnson.

"According to Group, we can expect a maximum effort against the Southeast. They estimate the window for an invasion is open until mid to late September. The summer is going to heat up some more, I'm afraid."

"I think we can all understand land versus water rescue," said 'Jackstay.' "That will give Gerry another 20 minutes of penetration. How can that be a good thing?"

Darling stared at Beamish, and then scanned the faces looking at him. "I do not want this even mentioned outside this tent." Their commander received various forms of confirmation. "What is not said but implied by this instruction is frankly our losses have been precipitous and cannot be sustained. Air Vice-Marshal Park is doing what must be done to preserve what resources he has left. That is the un-said meaning of this instruction."

"Is it really that bad, Skipper?" 'Fog' asked.

Darling scanned the faces of the pilots. "Yes. I am afraid so. We have had a rough go of it, but One One Group has been hit very hard. We are doing our part to help, when we are not protecting our sector, but One Two Group seems to be in a bit of a row with 'Stuffy' over this Big Wing tactic."

"I sure hope they get that debate sorted out promptly," Beamish said.

"Yes, as we all do. One One Group is dealing with minutes and seconds," Darling added. "They simply do not have the luxury of generating massed formations."

"What are those items about the Polish and Canadian squadrons?" asked 'Boxer' Stockard.

"Number One Canadian Squadron is still technically in training as well as the Polish Squadron. Committing these squadrons before they have completed their preparation is certainly an indication of how severe the situation is in the Southeast."

"Did they say anything about moving us back to Northolt?" 'Waggle' Davies asked for most of them.

"No, although we can be expected to fly more operations in support of One One Group."

"Hell of a deal," commented 'Red' Burns. "They chased us out of France. Are they going to chase us out of England, now?"

"Over my bloody dead body they will!" growled 'Sparky' Morrow.

"Here, here," added several others.

"What about the two fighters for the solo recce aircraft?"

"They were using sections as we have been, but resorted to singles several weeks ago. High altitude flight necessitates a companion. That is about the best I can say."

The telephone rang as the rain began its drumbeat on the tent. Corporal Warren answered the telephone and handed it to 'Spike' Darling. Their leader nodded his head several times, and then replaced the handset in its cradle.

"We have been released for the day. The weather is getting worse, and they do not anticipate any raids for the remainder of the day. They are trying to give as many squadrons as possible a rest period. Let's take advantage of the respite. Don't do anything foolish, and I expect everyone back here bright and early tomorrow morning except for Koenig. 'Slim,' you have a day pass."

"Thanks, Skipper," Koenig said with a touch of relief in his voice.

"The lorry should be outside to take us back to the Mess. Corporal Warren, please secure the tent."

"Yes, sir," she answered, as the pilots began to depart and run through the rain to the covered truck.

Darling waited in the rain, watched the truck move up the road, and then crossed the road to join his wife. The pilots ate a mid-day meal at the early sitting and scattered in numerous directions.

Jonathan grabbed Brian's elbow. "Tell me you are not going to London to see the Spencer woman."

"I thought about it but no. I'll call her to tell her I can't see her because we are so busy fighting a war."

"Didn't you try that once before, and didn't she come down here?"

"Yes, but I'll ask her not to come.  Maybe I can talk her into telling me her news on the telephone."

"Maybe," Jonathan said, as they both withdrew briefly into their thoughts.

Although Brian had not really considered calling Mary before he left to see Charlotte, he now wondered how he could talk to her?  She was always very persuasive.  As had happened in their previous encounters, Brian found it particularly difficult to resist her.  Jonathan was spot on, as he had become accustomed to saying.  If he did not meet her, eventually she would show up at the Officer's Mess making any resistance futile.  Unfortunately, he told himself, he could not deny the attraction he felt.  She was an impressive woman in many ways, and he did enjoy being around her strength.  Mary Spencer was a good woman even though she did lead him into an illicit affair.

It was Jonathan 'Harness' Kensington that returned first.  "So, then, what are you going to do?"

"'Jackstay' said he was not going to use his motorcar, again.  So, I asked to borrow it.  I want to apologize to Charlotte Palmer."

"Apologize for what?"

Brian stared at his friend for several moments, and then looked around not wanting other listeners.  He calmly recounted the events of his visit to Standing Oak Farm.

"Maybe she does not want to see you."

"Perhaps, but I can't leave this the way it is."

"Be respectful, Brian."

"I will," he said.  A much better idea came to him bringing a broad smile.  "Why don't you come with me?  Maybe she will feel less uncomfortable if you were there."

Jonathan laughed.  "You have so much to learn.  If she felt uncomfortable with one RAF pilot, how do you think she would feel with two RAF pilots bearing down on her?"

Brian felt a bit silly.  His best friend was correct as he usually was in cases like this.  "Just a thought," he protested.

"You go to apologize for your ungentlemanly behavior.  Maybe you can convince her you are just a brash young fighter pilot and a crude colonial," Jonathan laughed out the words.

"Bastard."

"Not really," he answered, and then released his smile.  "At least as far as I know," Jonathan said, returning his smile.

"What are you going to do?"

"I'll call Linda.  If she is available, I will head into London for an evening of sweet bliss."

"Excellent.  Please say hello for me.  Now, I think I will get going."

"Sure, and if you would say hello to Charlotte for me."

"Will do."

The pilots of No.609 Squadron, as well as sister squadrons at Middle Wallop, began to disappear from the Mess, each of them in the pursuit of their choice of relaxation, entertainment or pleasure, and push away the stress of mortal combat.

———

*Tuesday, 20.August.1940*
*Special Operations Executive*
*64 Baker Street*
*Marylebone, London, England*
*11:30 hours*

**O**nly the brass number identified the location of the five-story, rather ordinary, office building.  A non-descript restaurant and haberdasher store occupied the ground floor spaces.  As instructed, Trevor Andersen entered the simple entryway, ascended the only set of stairs to the first floor.  The entryway stairs did not go above the first floor.  The landing had passageways on both sides of the stairway, and yet there was only one door halfway down the right side passageway.  Again, as instructed, he did not knock and simply entered the small office.  A male receptionist sat behind a simple desk.  Only a portrait of the King behind the receptionist hung on the medium green walls.  The Union Jack hung off a single pole, upright in a corner stand.  Two chairs backed up to the wall on both sides of the door, facing the receptionist's desk.  There were no windows and only one other door to the right of the desk.

"May I be of assistance, sir?" the receptionist asked.

"I have an appointment at 11:30 hours."

"What is your name, sir?"

"Trevor Andersen."

The receptionist checked what appears to be a schedule.  Apparently satisfied, he said, "Please take a seat, Mister Andersen, if you would be so kind."  The man pushed an intercom button and announced, "Mister Andersen has arrived."  There was no discernible response.  "They will be with you, shortly, sir."

'They' . . . plural . . . Trevor wondered who the one or more others might be.  He had understood his interview would be with the minister.  Andersen did not have to wait long.  Trevor recognized Minister of Economic

Warfare Hugh Dalton entering from the far door.  He wore light brown suit with a plain blue necktie.  As he approached, he extended his hand.

"Thank you for coming over to our humble little sanctuary, Trevor.  I am Hugh Dalton."

"A pleasure to meet you, sir."

"The honor is mine," he said and smiled.  "Your reputation precedes you."

"Thank you, sir."

"If you would follow me," Dalton said and turned back to his entry door without waiting for a response.

The passed through another small room with two, heavily armed, rather severe looking guards dressed in Army infantry uniforms with no other designations or even rank.  Trevor followed Dalton down a blank corridor, ascended two more flights of interior stairs, and entered the last of four, closed doors on the left side of the stairway that continued to the upper floor.  They were in a functional conference room, with wall maps of the United Kingdom, Western Europe, Eastern Europe, the Mediterranean Sea, India, and Indochina.  There were several small blackboards, recently erased, scattered among the wall maps.  Another person entered after them.  Trevor recognized the man – Colonel Colin McVean Gubbins, DSO, MC, an outspoken and long-time advocate of irregular or unconventional warfare.  Gubbins had just recently been awarded his Distinguished Service Order for his leadership of the Independent Companies during the Norway Campaign last spring.  He wore a British Army service uniform with a Sam Browne belt across his chest and only his epaulette rank insignia – no unit insignia or personal award ribbons or insignia were present.

"Colonel, this is Trevor Andersen," Dalton said.

As Gubbins extended his right hand to Trevor, "Trevor, may I introduce Colonel Colin Gubbins, the newly appointed Director of Operations and Training at SOE."

"I have admired your work, Colonel," Andersen said.

"You have read my papers?"

"Yes, I have Colonel . . . all of them, I believe."

"You are one of the few, I must say."

"No need for modesty in here," Dalton added.  "There is a direct reason you have this job, Colonel."

"As you say, then, Minister."

"Let's begin.  Would either of you care for tea and biscuits?"

"No, thank you, sir," Gubbins and Andersen answered in unison.

"Then, let us be seated," Dalton said.  The two SOE executives sat across the table from Andersen.  "I understand you have discussed this assignment with Admiral Pike."

"Yes, sir, yesterday afternoon."

"We acknowledge that he was not particularly enthusiastic about losing your services.  Yet, both Colonel Gubbins and I are reasonably familiar with at least some of your work for Naval Intelligence and especially your operation in Pila, Poland a year ago winter."  Andersen did not react or respond with even the slight expression change and waited for the Minister to continue.  "Actions like Pila are precisely what SOE has been charged by the War Cabinet to carry out in all occupied territory.  Here, I will speak for myself.  Colonel Gubbins may wish to express his own opinion.  I think you are the prototypical field agent we need in SOE.  I understand from Sir Geoffrey that you are here because you see sufficient fit for your skills and experience."

"If I may, Sir Geoffrey recruited me into Naval Intelligence directly from Cambridge.  To be candid, he has been my mentor since then.  I am sure you can appreciate the loyalty I have for him."  Both Dalton and Gubbins nodded their agreement.  "I would be less than truthful and forthright if I did not confess my reluctance to leave Naval Intelligence as a consequence of my loyalty to Sir Geoffrey."

"Thank you for your honesty, Trevor.  Sir Geoffrey and I have discussed that matter thoroughly.  We have also agreed that you should maintain your relationship with Admiral Pike and Naval Intelligence, both professionally and personally.  Yet, I think you will quickly recognize, your skills are ideally suited to the SOE mission.  We would like you to join us.  Colonel Gubbins?"

"Minister Dalton has stated my perspective, precisely."

"Prime Minister Churchill succinctly stated our mission when he said, SOE must set Europe ablaze.  I believe he meant his statement literally and directly.  Our job will be to make connection with, encourage, supply and direct indigenous resistance movements throughout the occupied territories.  We will also conduct discreet sabotage operations as tasked by us," Dalton said, gesturing to Gubbins.  "To cut to the chase, is this an assignment for which you are willing to volunteer?"

"Yes."

"Very well, then, thank you and welcome to the team."

"Thank you, sir."

"I understand you come here with your recommendation regarding Colonel Pordonski."

"Yes, sir. Admiral Pike and I discussed Colonel Pordonski. I assume Sir Geoffrey has briefed you on Colonel Pordonski's service and my personal relationship with Stan."

"Yes, he has, thoroughly."

"I can assure you, Colonel Pordonski has extraordinary experience and skills, as well as extensive contacts in Eastern Europe."

"We have reviewed the debriefing notes, including the details of his exfiltration from Poland after the surrender," Gubbins said. "That experience alone qualifies him for service with us."

"That is my opinion, sir."

"Do you have any doubts, even the slightest question, about his trustworthiness or commitment to our cause?" asked Gubbins.

"None, sir. He holds no love for the Germans. I have no doubt he would be a dedicated operative in clandestine activities against the Germans."

"He is beyond the outer threshold of our age requirements."

"Put him to the test, Colonel. He will surprise you with his strength, endurance and stamina."

"Is he willing to do that?"

"I did not ask him that, but I know he wants to serve in the field, and knowing him, I believe he will be eager to demonstrate his capabilities."

Dalton did not engage. Gubbins considered Trevor's response. "Very well, then," Gubbins said, finally. "I understand he speaks English and German, in addition to his native Polish."

"Yes, sir, as well as passable Russian and Ukrainian."

"Excellent. We will make arrangements to speak to Colonel Pordonski tomorrow," responded Gubbins. "You are welcome to discuss this conversation with Colonel Pordonski, as you may choose."

"Thank you, sir. I will."

The men discussed the administrative details of Trevor's transfer from Naval Intelligence to the Special Operations Executive, with an expected effective date of Friday and a report date of the following Monday. They would take care of making all the arrangements to engage Colonel Pordonski and get him settled in London. Hugh Dalton would coordinate with his counterpart, the First Lord of the Admiralty, and the Director of Naval Intelligence to close the deal.

———

*Tuesday, 20.August.1940*
*Standing Oak Farm*
*Winchester, Hampshire, England*
*14:10 hours*

**B**rian had left RAF Middle Wallop, headed south on the A343 roadway, and worked his way toward Winchester and Standing Oak Farm, west of the city. The rain, occasionally heavy, had slowed his progress, taking him twice as long as the first time he drove to see her. He turned off the main carriageway at the simple but elegant Standing Oak Farm sign driving up the gravel road toward the farmhouse.

Brian stopped the car before the last rise. He asked himself several times if he was doing the right thing. He also rehearsed his words many more times. He had to avoid another rejection. He promised himself he would accept whatever boundaries Charlotte Palmer chose to impose in order to preserve some connection with his savior.

Without knowing why, Brian stepped out of the automobile in a light drizzle to walk the several yards toward the crest. He took his hat off so as not to be so obvious and peered over the crest to the farm buildings below. No one moved outside. There were no other vehicles evident. To his relief, warm light filled the windows of the main building. It looked like she was home. Brian took a very deep breath, returned to Roger Beamish's car and drove the remaining distance to the farmyard. As the gravel crunch and motor sounds announced his arrival, he saw her elegant face in the window. Her expression changed from curiosity to resentment as she recognized him.

Charlotte Palmer waited for him at the front door with her arms crossed under her chest and a stern, cold expression on her face. "You are not welcome here, Mister Drummond," she said through gritted teeth.

The words stabbed at his heart as he stood several yards away oblivious to the rain soaking his uniform. "Charlotte, I came unannounced because I felt you would not invite me."

"You are quite correct."

"I had to come. I must apologize to you properly," he paused, hoping for an indication he could come into the house. None came. "You saved my life, and I am deeply grateful for your courage. It would be tragic for me to lose contact with the woman who gave me a second life."

"I am not your mother, Mister Drummond," she said without the slightest possible crack in the rigidity of her position and tone.

Brian shuffled his feet and kicked a few rocks. "I didn't mean it that way. I never meant to offend or impose upon you."

"Yes, you did," she spat.

"I'm sure that is the way it must seem or appear, but I can assure you it was only my gratitude for what you did for me."

"That kiss was not a kiss of gratitude."

He shuffled his feet some more as he grappled with his feelings and thoughts. "To be absolutely honest and with the greatest candor I can muster up, I also find myself attracted to you."

"As I thought," she spat, again.

"Charlotte, you are a beautiful woman."

"I am also a widow grieving the loss of my husband to this damnable war, and you are an RAF fighter pilot who risks instant death in that damnable cockpit you love so much."

"An unfortunate price for freedom," he responded more instinctively and quickly than he should. She gave him no reaction. "I am terribly sorry for your loss, but fate brought us together." Without thinking, Brian lowered himself to his knees before her in the rain. The gravel cut into his knees. He clasped his hands together in front of him. "I most humbly seek your forgiveness for my inappropriate behavior." He remained frozen, as if he would not move until she forgave him.

Her expression did not break. She glared at him probably wondering what he was going to do next. The only break in the stalemate came when Charlotte stepped back and started to shut the door. Brian did not move other than plead with his eyes. She stopped, and then ever so slightly a smile began to form at the edges of her full lips. Warmth began to slowly fill her eyes. "You silly, sod. Come inside out of the rain," Charlotte said, standing sideways and motioning for him to enter.

Brian stood dripping on the stones of her entryway. He did not want to get her wooden floors and rugs wet.

"Come, now, you must get out of those wet clothes before you catch yourself a death of cold."

"I'll get everything wet."

"Well, then," she said, laughing and turning back toward him. "You have several choices. You can strip down where you stand. I shall not be embarrassed. Or, you can track a little water across the floors to the bath. Your preference, I'm afraid," she said, smiling broadly at him.

Brian considered her offer as maybe a means to place himself at her mercy. In the end, he decided not to press his luck with her. He had gotten this far, and he had no intention of losing the small ground gained.

"I'll take the bath, if you don't mind a little water on the floor."

She giggled. "I don't mind, as you say." She walked, and then motioned toward the larger than expected bath room. "I suggest you strip down promptly. I shall draw a warm bath for you while I search for some temporary clothing." She started the water in the large tub. Soon, steam began to rise from the accumulating water as he removed his tunic, tie, shoes, socks and shirt. He hesitated at his trousers and underwear. "There, nice and warm." She turned to look at him waiting on her. "Not to worry, Brian. Us farm girls are quite accustomed to anatomy. Now, why don't you take off the rest of your wet clothes, so I can dry them out." He did as he was requested as she stood nonchalantly watching him. He moved quickly to the hot water in the large tub. It took him several seconds to ease himself into the water. Charlotte had gathered up all his clothing. "I shall hang your uniform in the warmth of the kitchen. It will probably take several hours, at least, to dry. I am not so sure whether I will be able to find any clothing your size, but I shall try."

"Thank you."

"You are most welcome," she answered, as she started to leave, and then turned back to him. "By the way, you are forgiven for your transgression."

"Thank you, Charlotte. I am grateful to you once more."

"Nonsense. Now, just relax in that hot tub. I shall return shortly."

The hot water had precisely the desired effect. The chill left quickly, and the warmth extracted all the remaining tension. He even allowed himself to drift off into a light slumber. Her knock at the door brought him back and sat him upright.

"Come in," he answered.

Charlotte carried a large, thick, terry cloth bathrobe. "I am afraid this is the best I can do for the moment," she said, holding the robe. "You are much larger than either Ian or me."

"I don't mind."

"Smashing. Then, I shall leave this for you, when ready."

"Thank you."

"You are most welcome," she answered, as she turned and departed closing the door behind her.

Brian stayed in the bathtub until the water began to cool. He dried off, found a comb for his hair and donned the robe. It barely reached his knees, but seemed to cover him reasonably well. He joined her in the kitchen. His uniform items hung from hangers on various cabinet doors. The stones of the kitchen floors were darkened from the small puddles of water beneath each piece.

"I baked some scones and made some nice hot, white tea. I have taken the liberty of adding sugar."

"That's great.  I thought sugar was scarce."

"It is, but this is a special occasion."  She smiled.  "It is not every day a woman receives an apology for an innocent mistake from a man on his knees in harsh gravel and drenching rain," she said and laughed.

"It was heartfelt."

"I am sure it was," Charlotte said, as she motioned to a chair at the kitchen table.

They talked about many things through the afternoon as they managed to bridge the gap created by his *faux pas*.  Brian knew he had succeeded as he felt the warmth and connection with her.  He learned more about her deceased husband.  Brian added a few details from the military information they were provided.  He told her about David MacGregor, one of his mates at OTU7 last year, who also lost his life in the sinking of HMS *Glorious* by the guns of the German battlecruiser *Scharnhorst*.

They both tried to keep the conversation light and filled with laughter.  It made the rigors and demands of war remain in the distance somewhere.  It made them both feel better.  Brian forgot about everything beyond Standing Oak Farm.

"Isn't it milking time, soon?" asked Brian, not wanting to cause any unnecessary disturbance in the routine of the farm.

Charlotte Palmer looked at the wall clock adjoining the living room.  "Not quite yet, but it is about time for the Prime Minister's address."

"What address?"

"The BBC announced earlier today that the Prime Minister's speech in Commons this afternoon would be broadcast live.  Apparently, it is to be an important speech.  What do you think it could be?"

Brian churned through numerous thoughts about possible topics.  Maybe the war was going worse than anyone thought, and he was going to announce their surrender?  Or, maybe they had reached a temporary cease-fire with Germany?  Although that seemed very unlikely given the enormous bombing effort the Germans conducted in the last few weeks.  There were not many allies left, and he doubted the United States would have changed its position so quickly.  Maybe there was some other military tragedy like the sinking of the *Glorious*?

"I have no idea," he eventually answered, not wanting to concern her with the dark possibilities.

"Let me refresh your tea, and then we should go into the sitting room where the radio is."

As they settled into their chairs with the music from the BBC, Brian could not avoid the delicate, porcelain fine, features of her face.  Her prematurely gray hair shimmered like bright silver in the light of the fire burning in the large fireplace across the room.  As the clock struck eight bells, the nautical reference to 16:00 hours, the announcer prefaced Prime Minister Winston Churchill's speech to the House of Commons, to the nation, and to the world.

The shuffling sounds of the Commons coming to rest amid the muffled rumblings of the Members defined the live broadcast.  "Order, please," called Speaker of the House of Commons Edward Algernon FitzRoy, DL, Member of Parliament for Daventry.  He waited for silence.  "Mister Prime Minister, you have the floor."

"Almost a year has passed since the war began," came the characteristic speech of Winston Churchill filling the living room of Standing Oak Farm as Charlotte and Brian listened intently, "and, it is natural for us, I think, to pause on our journey at this milestone and survey the dark, wide field.  It is also useful to compare the first year of this second war against German aggression with its forerunner a quarter of a century ago.  Although this is in fact only a continuation of the last, very great differences in its character are apparent.  In the last war millions of men fought by hurling enormous masses of steel at one another.  'Men and shells' was the cry, and prodigious slaughter was the consequence.  In this war nothing of this kind has yet appeared.  It is a conflict of strategy, of organization, of technical apparatus, of science, mechanics and morale.  The British casualties in the first twelve months of the Great War amounted to 365,000.  In this war, I am thankful to say, British killed, wounded, prisoners and missing, including civilians, do not exceed 92,000, and of these a large proportion are alive as prisoners of war.  Looking more widely around, one may say that throughout all Europe for one man killed or wounded in the first year perhaps five were killed or wounded in 1914-15."

"That is not a particularly comforting thought," remarked Charlotte.

"No, but it is important."

They listened as Churchill tried to find brightness in their present situation.  Brian knew more intimately the tenuous hold they maintained on freedom.  They exchanged eye contact.  He did not want anything to happen to her or any others close to him.  Somehow, they had to prevail, to defend freedom and eventually repel the aggressor.  His attention returned to the radio.

"Hitler is now sprawled over Europe.  Our offensive springs are being slowly compressed, and we must resolutely and methodically prepare ourselves for the campaigns of 1941 and 1942.  Two or three years are not a long time, even in our short, precarious lives.  They are nothing in the history of the nation, and when we are doing the finest thing in the world, and have the honor to be in sole champion of the liberties of all Europe, we must not grudge these years or weary as we toil and struggle through them.  It does not follow that energies in future years will be exclusively confined to defending ourselves and our possessions. Many opportunities may lie open to amphibious power, and we must be ready to take advantage of them. One of the ways to bring this war to a speedy end is to convince the enemy, not by words, but by deeds, that we have both the will and the means not only to go on indefinitely but to strike heavy and unexpected blows. The road to victory may not be so long as we expect. But we have no right to count upon this. Be it long or short, rough or smooth, we mean to reach our journey's end."

The cheers of the House of Commons punctuated his speech. They seemed to need his encouragement as much as the people and the warriors. Charlotte looked to Brian with questions in her eyes. Brian could only smile and nod his head in recognition and partial reassurance.

"As I explained to the House in the middle of June, the stronger our Army at home, the larger must the invading expedition be, and the larger the invading expedition, the less difficult will be the task of the Navy in detecting its assembly and in intercepting and destroying it on passage, and the greater also would be the difficulty of feeding and supplying the invaders if ever they landed, in the teeth of continuous naval and air attack on their communications. All this is classical and venerable doctrine. As in Nelson's day, the maxim holds, 'Our first line of defense is the enemy's ports.' Now air reconnaissance and photography have brought to an old principle a new and potent aid."

More cheers came across the airways. This speech made Brian's fatigue disappear, his concerns, his worry about what lay ahead vanished among the melodic words. He felt good. A voice within him told so much more about the man behind the words.

"The gratitude of every home in our Island, in our Empire, and indeed throughout the world, except in the abodes of the guilty, goes out to the British airmen who, undaunted by odds, unwearied in their constant challenge and mortal danger, are turning the tide of the world war by their prowess and by their devotion. Never in the field of human conflict was so much owed by so many to so few. All hearts go out to the fighter pilots, whose brilliant actions we see with our own eyes day after day; but we must never forget that all the time, night after night, month after month, our bomber squadrons travel far into Germany, find their targets in the darkness by the highest navigational skill, aim their attacks, often under the heaviest fire, often with serious loss, with deliberate careful discrimination, and inflict shattering blows upon the whole of the technical and war-making structure of the Nazi power. On no part of the Royal Air Force does the weight of the war fall more heavily than on the daylight bombers who will play an invaluable part in the case of invasion and whose unflinching zeal it has been necessary in the meanwhile on numerous occasions to restrain."

Brian swelled with pride.  Charlotte looked at him with tears in her eyes and an odd expression of admiration, curiosity and maybe even a little worship.  He was a part of something much bigger, more profound, more significant and more deterministic than he had ever imagined listening to Malcolm's stories of flying with the Royal Flying Corps during the Great War.  For the first time, Brian felt the mantle of history enveloping him and his brother warriors.  He also felt success within the darkness.  He began to feel uncomfortable as Charlotte continued to gaze at him like he was an exotic zoo animal or vaporous apparition.

"Presently we learned that anxiety was also felt in the United States about the air and naval defense of their Atlantic seaboard, and President Roosevelt has recently made it clear that he would like to discuss with us, and with the Dominion of Canada and with Newfoundland, the development of American naval and air facilities in Newfoundland and the West Indies.  There is, of course, no question of any transference of sovereignty – that has never been suggested – or of any action being taken, without the consent or against the wishes of the various Colonies concerned, but for our part, His Majesty's Government is entirely willing to accord defense facilities to the United States on a 99 years leasehold basis, and we feel sure that our interests no less than theirs, and the interest of the Colonies themselves and of Canada and New-foundland will be served thereby.  These are important steps.  Undoubtedly this process means that these two great organizations of the English-speaking democracies, the British Empire and the United States, will have to be somewhat mixed up together in some of their affairs for mutual and general advantage.  For my own part, looking out upon the future, I do not view the process with any misgivings.  I could not stop it if I wished; no one can stop it.  Like the Mississippi, it just keeps rolling along.  Let it roll.  Let it roll on full flood, inexorably, irresistible, benignant, to broader lands and better days."

The cheers and clapping on the radio joined the joyful clapping of Charlotte Palmer.  Tears streamed unabashedly down her cheeks.  Brian felt the tears in his eyes, which he quickly wiped away.  The calm strength of Churchill's words complemented his careful, firm, unwavering delivery.

"Oh, dear," Charlotte said eventually.  "Look at me, weeping like a school girl."  She dabbed her tears with a small, lacy handkerchief.

"I felt it, too."

"The few he said."

Brian nodded.

"Is it really like that?  I mean it sounds like we are so close to the edge of the abyss, and it is you and your mates keeping us from falling into the crevasse."

"I don't know if it is quite like that, but we are seriously outnumbered in the air."

"Are they going to beat us?  Are they going to invade?"

"Not if we have anything to say about it.  We have inflicted rather heavy losses on the enemy with our small numbers.  The Germans do not control the skies.  We do."

Charlotte glanced at the wall clock and stood quickly.  "Look at the time.  My poor cows must be beside themselves.  I'm afraid I must ask you once more to help me, if you would be so kind?"

As she hurried about gathering a coat and oilskin hat, Brian considered whether he should get back into his still damp clothes.  In the end, he decided to put on his shoes without socks and join her in the barn.  It was still raining as he ran to the barn.  He had to be quite a sight in a relatively short, white bathrobe and black shoes with nothing else on.  He found the appropriate pail and stool, and went to the second cow.  The animals were groaning as a result of their distended udders.  He moved quickly as she had taught him.  It was not until they were both on their next cow that either of them spoke.

"Brian, thank you for helping me milk the cows."

"My pleasure."

"I did not realize how much time passed during the Prime Minister's speech."

"No problem.  We got to them before they burst."

"I must say, I do like your outfit.  Quite stylish, I'd say," she said with a strong laugh.

"Maybe I'll start a new trend."

"Maybe."

They finished the task, released the cows and cleaned their tools.  Before they left the barn, Charlotte stood in front of him.  She looked deep into his eyes.  Brian felt an electricity pass from her wondrous blue-gray eyes.

"I must apologize to you," she said with solemnity.

"What for?"

She broke eye contact with Brian.  It was her time to shuffle her feet among the loose straw on the barn floor.  "I want to apologize for my overreaction to your kiss.  You caught me by surprise, and I just reacted.  You are a very nice young man.  You have been quite helpful with the cows.  I certainly appreciate your gratitude, but please do not feel indebted to me.  I had no choice. I simply did what had to be done. Truth be told, as I have said before, I had given up, and it was your mates that kept me at the task."

"You have no apology to make. I was out of line, and it is me who must apologize. I want to maintain contact with you. I must maintain contact with you. It is very important to me. I hope you understand."

"I think I do."

"Now, I think I should go."

Charlotte locked on his eyes. She searched for something deep inside him, something he could not see or feel himself. Charlotte Palmer leaned forward and kissed him on the cheek, and then turned. They walked silently back to the house.

Brian grappled with his confusion and conflict. He did not know which way to turn at this fog bound intersection. He wanted to stay. He desired to be close to her, much closer to her, but for the first time in his life he actually feared losing someone or something. Although he wanted intimacy with this intriguing woman, he was willing to forego any physical-ness just to be near her. She saved his life, gave him new life, and represented an entirely different dimension to life.

She never looked back to see if he was following her. Maybe she could just sense his presence. The crunch of gravel and the door eventually closing behind her confirmed his location, but no words passed between them. Charlotte went directly to the kitchen and slowly, gracefully and methodically felt his clothes. Without looking at him, she said, "They are still damp."

Brian considered for a moment whether that observation might be an invitation to stay. The moment of consideration was brief. "I'll be OK," he answered, as he moved to gather up his clothes. She looked over her shoulder at him. "I appreciate your hospitality, and I thank you so much for your forgiveness." She nodded her head, and then turned her head away. "You represent something very special to me. I hope you can understand why I need to be close to you, and I hope you will let me."

"I think I understand."

Brian finished gathering up his underwear, socks, shirt, tie and trousers. He would leave his tunic and hat for his departure.

She turned to him. "Your clothes are still damp. Why don't you at least stay until they are dry?"

"That may be half the night."

Charlotte averted her eyes. Brian felt his heart pounding as the electric charge built rapidly within him. He sensed what she wanted, or maybe what he wanted her to want.

She smiled and fidgeted. "Please stay."

Brian knew this was a critical juncture, and yet the paths forward remained shrouded in fog. He could not see, think, sense or feel a clear direction. The relationship with this woman had to be more than sex. While he could feel her heat and he wanted the comfort of her warmth, the risk seemed too great. "Thank you, Charlotte, but I must get back. I've borrowed my section leader's motorcar, and my squadron leader wants us ready to fly first thing in the morning."

She bounced forward toward the stove like restraining bindings had released the spring. "Yes, of course. I shall make some fresh tea before you go."

Pilot Officer Brian Drummond dressed slowly as he eased himself into the cool, damp uniform. She was seated at the kitchen table. She poured his tea, added the requisite amount of sugar and milk, and placed his cup of tea at the chair across from her. He sat down and looked at her perfect face, eyes, nose and lips. Charlotte Palmer was a simple, but incredibly attractive woman. It would be so easy for him to stay with her. He wanted to stay with her, but he did not trust himself to make the correct choice. Better not to choose at all.

They returned to the lighter side of life, sharing the little things that seemed so funny to them. They could laugh about the odd image of Brian kneeling in the rain, the oddities of the farm, and the relief of the rain. Brian liked forgetting about the war and the mortal struggle for aerial supremacy.

Brian departed in a light drizzle. He could see her in the rearview mirror, standing in the doorway as he drove away. She did not move until he could no longer see her. The drive back to RAF Middle Wallop did not take long in his mind as he thought of the afternoon's events, the magnetic tension between them and the confusion of his feelings. He shouted loudly several times, as he drove the distance, as if to relieve excessive pressure in a steam boiler. Brian convinced himself he had to take things more slowly with Charlotte Palmer. Another mistake, and he might lose her.

---

*Wednesday, 21.August.1940*
*RAF Middle Wallop*
*Middle Wallop, Hampshire, England*

The fog thickened substantially during the night that made the waiting considerably less stressful. They all knew they would not fly in these weather conditions. The meteorologists in the Middle Wallop Meteorological Office predicted the weather would remain for another day or two. None of them seemed to mind that they were required to be at their dispersal tent at Available status.

The discussion centered around one topic, Churchill's speech to the House of Commons yesterday afternoon. Brian did not participate much in the conversation as his mind ground away at the block of granite Charlotte Palmer seemed to represent. Everyone boiled the speech down to that one sentence. The ground crews, civilians and even Charlotte Palmer focused on the moniker – the few. They also managed a few jokes, as the pilots often did, with variations of Churchill's reference. They added, '. . . and paid so little.' None of them objected to the recognition by the Prime Minister, but none of them had the reverence others appeared to display almost instantly. They shared stories of what they were doing when they heard the speech, or if they did not hear it, when they were told about it. The moment in time with Charlotte Palmer proved too personal for Brian, and he could not add it to the verbal log.

The impact of the Prime Minister's speech changed the mood in and around everyone. While just a few days ago, the mood was stern, quiet, resolute determination to succeed; now, there was effervescence, and ebullience to the words, actions and appearance of people. Their backgrounds or professions did not seem to matter. A detectable air of optimism lifted everyone's spirits. The respite offered by Mother Nature allowed the mood to germinate and grow. While the pilots knew they had many more days, weeks and months of mortal combat ahead, for the first time they began to believe, actually believe, they could and would succeed.

Brian tried to participate just enough to not draw attention to his contemplative mood. The majority of his thoughts rested unquestionably on Charlotte Palmer. The attraction pulled him closer, but his fear of offending her and losing his connection with her dominated his cogitation. Yesterday, he thought he could feel an attraction coming from her. He just hoped it was not wishful thinking on his part.

By mid-morning, the fog receded sufficiently for them to see the Operations building control tower from their dispersal location, but it was a long way from acceptable flying weather or ground conditions. The skies remained silent. Not even the German reconnaissance aircraft wanted to fly in this soup. Eventually, Group conceded that the weather would not allow defensive operations, and there were probably no intruders for them to defend against. The squadron was released just before lunch. They ate together including Squadron Leader Darling, and then began to split to enjoy their restive pursuits. Mansek, Kradilcek, Burns and Koenig departed promptly for London and Shepherds Pub among other haunts. Jonathan Kensington tried to coax his best friend into going to London as well, however Brian had other ideas. Within an hour of lunch, all the No.609 Squadron pilots were gone except for Brian.

Charlotte Palmer pulled him toward Standing Oak Farm, but prudence and caution kept him from the journey.  Confusion tore at his judgment.  He did not want to press any possible relationship with her.  He wanted some sign from her although he recognized he might never get such a sign.  Brian decided he would cross that road when he came to it.  He chose to refocus his thoughts on something else.  He would write a letter to his parents, and then maybe call Mary Spencer to see if he could persuade her to tell him her news over the telephone.

Brian went to his small room, pulled back the curtain to see the grayish tint to the green countryside defining the periphery of the Salisbury Plain.  The fog was gradually being replaced by more rain.  He sat at his diminutive desk looking out the window and pulled several sheets of paper from the drawer.  Before he began writing, he reread the letter from his parents.

---

*21 August 1940*
*RAF Middle Wallop, Hampshire*
*Dear Mom and Dad,*

*I received your letter of July 14, just yesterday.  I cannot tell you how important your letters are to me.  Please pass a big hello to Mrs. Bainbridge.  I wish there was something I could do for her.  I know she misses Malcolm more than me, so that makes her loss pretty bad.  I was glad to hear about Becky.  I know you liked her.  I think I loved her, Mom.  My first love I suppose, next to you of course.  I guess she had a real hard time accepting what I had to do.*

*Things here are OK.  It is raining today, so the squadron had been released from alert duty.  Most of my friends have gone into London, but I wanted to write to you.  Somedays things are easy like today, but other days are so intense you cannot imagine.  We fly so much and we get so tired sometimes it is hard to stay awake.  The country is preparing for the Nazi invasion, but we are going to keep them out.  They have thrown everything they have at us and we've kept them out.  We are going to win the Battle of Britain.  I just know it.*

*Yesterday, we heard Mr. Churchill's speech about us.  We liked the phrase he used, but several of the guys changed it slightly.*

*Never have so many owed so much to so few for so little pay.*

*His speech seemed to inspire everyone.  I really enjoy listening to him.  I am so glad that Group Captain Spencer, Malcolm's friend, took me to meet Mr. Churchill before the war.  I will never forget it.*

*I am all right.  You can rest assured I am doing everything I can to take care of myself.  So many things have happened since I have been in Great Britain.  I want to tell you all about it, but it will have to wait until this war is won and I return home.*

*Say Hi to everyone.  Keep those letters coming.  I love you and miss you both very much.  Please take care of yourselves.*

*Your loving son,*

*Brian*

---

Brian reread his letter several times and chuckled to himself each time he read the pilot's version of Churchill's words.  He also thought of Becky and their first steps toward the intimacy of a man and a woman.  He did not blame her for not wanting to stay with him.  He was as much to blame as she was.  She never had a choice when it came to his flying and his dream.  He left for England without even saying good-bye, and he regretted that part.  Brian found comfort in the fact that his parents were safe, in good health and at least beginning to accept his endeavor.  He folded the paper, slipped it into an envelope, addressed it and left it on his desk.  He would take it down stairs to the duty clerk who would in turn make sure it was mailed properly.

His thoughts turned to Charlotte Palmer, and then to Mary Spencer.  He needed to telephone her as she requested.  His curiosity about her news did occupy a fair portion of his thoughts.  He found her letter, reread it several times, and then searched for and found the Spencer's home telephone number.

Several telephones were provided to the officers in the sitting room of the Mess.  Only one telephone was placed in a small room that provided some privacy.  It was being used by one of the No.238 Squadron pilots.  He waited for ten minutes until the private telephone was available.  Brian closed the door behind him, sat at the small table, took a deep breath and called Mary.

"Bushey Heath Two Four Seven One," she said with her delightful voice.

"Mary, this is Brian."

Silent words and the soft crackle of telephone line static filled the distance between them.  At first, Brian thought she might be mad at him.  Then, he wondered if she was even on the other end of the electronic connection.

"Are you there?" he asked.

Again, silence, and then she responded with a muffled, silently sobbing voice.  "Yes.  I am here."

"What's wrong?  Are you OK?"

"Yes, yes," she answered with more pronounced choked back tears. "It was just that I did not think I was going to hear from you."

"Why?"

"I had a nightmare last night."

"But, it was just a dream," he said, as the memory of his nightmare came back to him.

"No, Brian," she said, crying more in relief than pain or hurt. "This was much more."

"It was still a dream."

"Then, you tell me. This dream seemed to be in color, vivid color. I watched you. It was so clearly you in the cockpit of your aeroplane and flames all around. I watched you struggle to get out, but you were trapped. I could see the pain and fear on your face, Brian. It was the most grotesque bloody nightmare I have ever had."

It was Brian's turn to be silent as he reeled under the weight of her words. Dreams did not happen like that as far as he knew. He could not avoid the comparison with his nightmare of two months earlier about being trapped in the telephone booth with bombed, broken, burning debris surrounding the small space. The connection gave him a deep chill. They were both dreams. They had no other value, purpose or objective. They were dreams, imaginations, maybe even hallucinations, but they were still just bad dreams.

He wanted to tell her about his nightmare before Anne's execution, but he knew if he drew the linkage it would rattle them both. This had to be expunged from his mind. Brian could not let these thoughts remain, or they would surely dull his edge in the air.

"Now, it is my turn to ask if you are still there?" Mary asked with more control in her voice.

Brian shook his head, as if he were shaking off a strong head blow. "Sure," he said as confidently as he could. "I was just listening to you. What can I do to help? I am perfectly fine. Nothing has happened to me, and I'm going to do everything I can to keep it that way."

"I want to see you Brian. I need to see you."

"I know. You said that in your letter, but we have a war on, you know."

"Funny. How come both you and John keep telling me that trite little phrase?"

"I'm sorry. I didn't mean it that way. It's just that we are either in the air or on alert."

"So, are you on alert now?  The middle of the day and week, a dreadful storm and none of those damnable, droning German bombers . . . does this mean you are on alert?"

"No, but we'll probably be back on standby first thing in the morning."

"Look, Brian," Mary said, now with a more ragged edge to her voice. "I truly do not want to interfere in your life, cause you any distraction, or make your life more difficult, but I simply must see you."

Brian's thoughts filled with the contradiction.  She had interfered in his life, caused him distraction and made his life much more difficult. "Can't you tell me your news over the telephone?"

"If I didn't know better, I would surely swear you do not want to see me."

"No.  It is not that at all," he lied.  He knew he did not want to see her alone.

"Then, what is it?"

Brian looked out the window in the door to see if anyone might be waiting, or close by.  The sitting room was empty.  The stale, smoky, stifling air in the small telephone room choked off his oxygen.  He tried to take a deep breath, but could not accomplish it.

"Time.  It takes so much time to get to London," he said with thoughts of his fellow pilots probably already in London.  "I don't have a pass, and our squadron leader needs us back in the saddle early in the morning."

"Then, I'll come to you."  Papers shuffled in the background.

"Can't you tell me on the phone."

"No!" she said rather sharply.  "This is too important and too personal."

"OK."

"It will only take me a few hours to get to Andover.  Please meet me at the rail station.  According to the rail schedule, I can be in Andover by quarter past four.  Please, Brian.  It is important to me . . . and, I think to you."

"OK."

"Bless you.  I shall see your precious face in a few hours then."

"OK."

"I love you, Brian.  See you soon," she said and hung up before he could say anything.

Brian sat in the stagnant air of the telephone booth thinking about what he had just agreed to and what he would rather be doing.  He wanted to see Charlotte Palmer, but he sensed it was too soon.  A knock on the glass brought him back to the RAF Middle Wallop Officer's Mess.  One of the No.604 Squadron night fighter pilots needed the telephone.

"Girl trouble, ay, mate?"

"I suppose," Brian answered as he vacated the small room.

"Hang in there, old boy. It'll pass. It always does."

"Right."

Brian asked the mess clerk to find out if any vehicles were making the relative short journey up the A343 to Andover. By the time he straightened his tie and retrieved his tunic and hat, the clerk indicated that a supply truck would stop by the Mess in a few minutes on its daily run to Andover. There was a seat for him. Brian stepped outside the door. The rain had stopped. A moderate wind churned the damp, earthy air.

He was deposited at the rail station in less than an hour, well ahead of Mary Spencer's expected arrival. Brian checked with the stationmaster to see if the trains were running on time. They were. He wondered whether he should try to find a room, but quickly discarded that notion. He did not want to ask for more trouble. He reread her letter and sensed they would ultimately end up in a bed somewhere, but he did not want to help the process along at all. She had never been particularly bashful about what she wanted from him. Brian walked down the nearby High Street. He blindly looked in the shop windows, nodded to the greetings of passing citizens and tried to eliminate the time between them. Several older men stopped to talk to him about the war, their experience, and then about his accent and citizenship. Although he was not in a talkative mood, he acceded to their requests. It was important to reassure the public just as the Prime Minister had done yesterday afternoon. As he desired, the time passed quickly. He returned to the station ten minutes early.

The train from London arrived as close to on time as trains ever did. The smoke and steam were slow to dissipate in the heavy, damp air. Several passengers disembarked as others boarded for their journey to the West country and probably Bristol or Cardiff. Mary Spencer appeared like a vision from the cloud of steam trailing back from the locomotive. A smile illuminated her face when she recognized the tall, trim, handsome RAF officer. Her pace quickened not quite to a jog toward him as he walked toward her. He rendered a smile although his heart beat faster with more than a little trepidation.

He expected a hug and kiss when she extended her right hand. He took her hand and touched both cheeks in the European fashion. She carried a small bag that he took from her.

"It is so good to see you, Brian."

"Good to see you, Mary."

"Let's walk," she said, motioning toward the station exit.  Mary Spencer took his arm, as they walked toward the town center.  "Have you found us a room?"

Brian had always been impressed by Mary's strength and confidence. She reminded him of Rosemary Kensington in many ways although more mature and refined, like a cross between Rosemary's brashness and Anne's sophistication.

"I didn't know how long you were planning to stay.  I wasn't sure if you just want to talk face to face, and then return to London."

Mary stopped, faced him and placed her soft right hand on his cheek. "Silly boy.  I would never miss an opportunity to be with you.  No one knows us, or at least no one knows me, in this village."

Brian knew what she had on her mind and wanted to attempt a deflection.  "What did you want to talk about?"

"Later, my dear, later.  I want the mood to be just right."

"You won't even give me a hint to prepare me?"

"An accomplished warrior such as yourself," she chuckled.  "You are quite used to responding to any situation."

They walked, talked about unimportant things and tried several inns until they found one that satisfied Mary.  The innkeeper, an elderly man, asked no questions and gave no indication of any reservations about his newest patrons.  Once in the room with the door closed behind them, Brian placed Mary's bag on the small dresser.  She moved slowly toward him, wrapped her arms around his waist and pressed herself to him.  She waited before she kissed him.  The kiss quickly became more involved.

"It is so delightful to feel your embrace."

"Thanks."

She lay her face upon his chest.  "I can hear your heart pounding.  Are you nervous?"

"A little."

"You needn't be.  I would never hurt you, and I will certainly protect you."

Brian nodded his head but did not answer her.  Although the risks kept him from being comfortable with her, Brian did appreciate the contrasts her elegance represented.  Brian's curiosity began to corrode his patience.

"What was so important?  What did you want to talk to me about?"

"Not now."

"What do you want from me, Mary?"

Mary Spencer pulled back from him only slightly. She reached for him and felt his response. "Pardon me for being so direct, but this is what I want."

Their intimacy progressed slowly. Brian relaxed and focused his attention on her. They enjoyed several summits together before they stopped. Mary lay against Brian, wrapped in his arms like spoons as they allowed the warmth of their intimacy to ripen the moment. They both drifted into a quiet, entwined and peaceful slumber of satisfaction.

When they awoke, Mary rolled toward him and kissed him several times, and then sat up in bed to face him. He enjoyed and appreciated the exquisite curves of her body.

"Brian," she said, and waited for his attention and eyes. "There is probably no delicate way to say this, so . . . I shall say it directly." She took a deep breath. "I am pregnant."

Brian smiled, raised himself onto his left elbow and touched her shoulder with his right hand. "That's great. Congratulations," he said, and then his expression turned serious. "I hope I didn't hurt you."

She chuckled. "No. You did not hurt me. You have made me feel very much the woman and quite content actually."

The image of the mechanics came to Brian bringing a new concern. "I hope we didn't hurt the baby."

Mary laughed in an animated, delightful way as if she was being tickled. "Brian, men and women have been doing this for centuries, for millennia. We can enjoy ourselves for many more months without hurting me or the baby."

"I'll bet Group Captain Spencer is excited."

"No."

"Why?"

"He does not know yet."

"Why?"

"The baby is ours, Brian."

The pendulum swung in a flash from elation to the fear of a caged animal. A cold, chilling blanket fell on him in that room. Brian felt his world collapsing around him. Everything he worked for turned to liquid in his hands. Strangely, he also felt a new, unknown glimmer of warmth & light among the cold of that moment. The thought of being a part of the creation of new life had intrigued him for a long time although he had never considered the opportunity to be so close. While there was certainly reason to be happy, it was the immediate threat of losing his Spitfire seat that occupied the majority of his conscious thought. He had impregnated the wife of his most important supporter.

Mary sensed the turmoil within Brian's head. "It is quite all right, Brian. Not to worry. I did not tell you this glorious news to cause you any sense of pain, anguish or obligation."

"But, I do. I must." Warmth began to return to his body and soul.

"Feel pain?"

"No, no. Obligation."

"No, Brian," she said sternly, taking on a far more serious expression. "That is not what I want."

"But, while the news and thought took me a little by surprise, I am the father. While I admit I didn't think it would be so soon, I kinda like the idea."

Mary sat up very straight with her legs crossed in front of her without the slightest flicker of self-consciousness or embarrassment with her nakedness. "Brian, listen to me. This is very important. While this baby, for me, represents a unique and special bond between us, it cannot alter our visible relationship."

"What do you mean? I thought you told me before, you and your husband had not been intimate in more than a year."

"I did tell you that because it was the truth, but I am sure you can appreciate that as soon as I suspected that I was pregnant, I had to take the appropriate steps." A curious, inquisitive expression occupied Brian's face. "I went to Bentley Priory and found a way to make love to John. No one will know this child," she said rubbing her lower abdomen, "is ours, except you and me."

"But . . ."

"No, buts, Brian. It simply must be this way. Anything else would hurt John, ruin your life and mine. It must remain our special secret."

Brian began to see the wisdom in her words, but found himself looking for a different path. He felt a new closeness with Mary Spencer, beyond the flesh of their previous relationship. Brian laid his head upon her right leg. She stroked his head running her fingers like a comb through his hair. A fascination with the growing life within her, a life he helped to create, began to grow with each moment, each thought and each image. He wanted to feel the life growing within her. It was the trust he had for her that eased his concerns and fears for the consequences. He trusted Mary Spencer as much as he did her husband.

"Brian," she said softly with her seductive, magnetic intonation. He nodded his head to indicate he was listening. "I want you to promise me something."

"What?" he asked, as he rolled slightly to look into her eyes.

"I want you to promise me you will act as if nothing has ever happened between us, and especially regarding this baby. It is extremely important. You must go on with your life, as if nothing whatsoever has happened between us. You must promise me you will act genuinely surprised when John eventually tells you of my pregnancy. You must find the right woman, marry her and have your own family. Promise me."

Brian hesitated, as if he needed time to consider the consequences of her request. "I promise," he said with realization and understanding.

"We can enjoy this when we are together, when we are alone, but we must never, ever let on to this link between us."

"I understand."

"Good," she said, as she repositioned both Brian and herself. "Now, make love to me again. Make love to me as you have never made love to a woman before. Let's enjoy the sins of our flesh and pleasure, before you must return to your beloved cockpit."

They wriggled and moved together with new vigor and enthusiasm. Brian did indeed feel a new closeness to this woman whom he knew would always be beyond his reach, as they joined and began to feel the rhythm of life.

Mary looked up at Brian suspended above her. "I wish you loved this cock pit," she said, emphasizing the words, "as much as that other one." She tickled the sides of his ribs, as she gripped him holding him close to her.

"Mary," he fained protestation.

"I know, that was rather raunchy. Wasn't it?"

"Yes, it was," he said softly, as he returned to her pleasure.

"I have always thought it rather spicy when lovers talk nasty in the throes of passion."

Brian found pleasure in her movements. The smile in her eyes and across her full lips added to his enjoyment. He adjusted his position to focus on Mary.

"A little higher."

Brian moved in response. She pressed against him, moving her hips to complement his efforts. "That's it. Yes, sweet heaven, you are good."

They enjoyed themselves, shared their thoughts and dreamed of the future. Brian knew he could not spend the night with her and be ready for operations early tomorrow morning. Mary understood the requirements of their defense. She received from him the assurance and love she wanted, and she felt the relief of a burden lessened. She finally had someone important to her to share her experience, concerns, worries and enjoyment; that was sufficient for her. They shared some bread and cheese the proprietor prepared for them along with an excellent bottle of cabernet sauvignon. Brian reluctantly left her in Andover and returned to his mission.

———

# Chapter 7

He told me never to sell the bear's skin
before one has killed the beast.

-- Jean de La Fontaine

*Thursday, 22.August.1940*
*RAF Middle Wallop*
*Middle Wallop, Hampshire, England*

## Week 7

The news Mary gave Brian brought a fresh, unique confusion. There was no one other than Mary he could share this with. Even Jonathan would go crazy and would probably not understand. Anne Booth would have understood, and been a good listener and objective friend, but she was gone. In the end, Brian decided he needed time . . . time to think, time to consider, time to refocus.

The rain had gone, but the low clouds remained. Unfortunately, the cloud cover did not stop the Germans. They launched on their first intercept in late morning, just before lunch was to be served. They hunted among the layers of clouds and found nothing.

Twice more during the afternoon, the pilots of No.609 Squadron launched against detected but unseen enemy raids. The clouds were dissipating, but the hunt still proved to be too difficult. The only satisfaction they found was the weather affected the Germans as much as it did the British. They heard no reports of bombs falling through the clouds.

By the time they reached the evening and retired to the bar after supper, they were ready for the relaxation of alcohol, camaraderie and war stories. The day's flying frustrations kept his mind off his personal confusion. The evening events helped as well.

Pilot Officer 'Boxer' Stockard joined them in the Officer's Mess bar a little later than the others. He looked like he had something he wanted to say and did not waste time getting to it. "I just heard from a good friend of mine who is flying in the southeast. He said the Germans fired long range, heavy artillery at Dover."

"From where?" asked 'Slim' Koenig.

"Cape Gris Nez, from what he said."

"Friggin' bastards can't beat us in the air or bomb us out, so they figure they'll use the anonymity of those friggin' cannons," barked 'Waggle' Davies.

"How far is that?" asked 'Red' Burns.

"I often forget you colonials may not know the significance of that little body of water," joked 'Sparky' Morrow. "It is 20 miles plus a little."

"Artillery can fire that far?"

"And some, I'm afraid."

"That ought to tweak old Winston's nose," offered Davies. "It should be interesting to see what he decides to do answering the Germans. Bombers are one thing, but artillery doesn't mind the weather and can go on around the clock."

"Not a particularly pleasant thought," added 'Harness' Kensington.

Brian went to the bar for a fresh pint of bitter. Near the end of the bar, Roger Beamish sat with a smaller, trim, blond haired man. Brian nodded to his section leader.

"'Hunter,' come on over here laddie," Beamish said motioning for him to join them. "I want you to meet the famous 'Cat's Eyes' Cunningham."

"Roger, you know I do not like that name."

"Oh, dear, and I like 'Jackstay.' The name fits, John, the name fits." Beamish turned to Brian and motioned toward Cunningham. "Brian, I am proud to introduce you to Flight Lieutenant John Cunningham, affectionately referred to as 'Cat's Eyes' in deference to his demonstrated prowess as a nighttime fighter pilot."

Brian extended his hand that Cunningham shook firmly. Cunningham's cool blue eyes held the quiet, skilled confidence of a stalking cat.

"And, you must be the American they call, 'Hunter,'" said Cunningham with refined courtesy and respect.

"Yes, sir."

"This young lad came to us fresh from the farmland of Kansas and has become quite an accomplished fighter pilot," Beamish added.

"So, I hear."

"Brian has been quite the student of air defense techniques. I am quite certain he would love to sponge up as much of night fighter operations as you would tolerate, John. What do you say, then?" Cunningham nodded his head with a smile. "Then, I shall leave you to it."

"First," said Cunningham, "I would prefer you call me, John. I rather abhor that moniker the lads have given me."

"No problem."

"Second, I am not much of a talker. I rather believe the adage, actions talk louder than words, so I'm afraid I shall have to ask you to provide the questions, which I shall endeavor to answer as best I am able."

"Agreed."

"Then, the floor is yours, as they say in Parliament."

Brian considered the information he had acquired over time, listening to other pilots talk about the night intercept problem.  Finding an elusive target in the ink black of the night sky and especially with the air raid blackened land beneath them had to be as bad as the stories.  Brian knew they were experimenting with a smaller version of the massive Radio Direction Finding equipment that made up the Chain Home air defense early warning system.  He also knew 'Cat's Eyes' Cunningham would not be able to talk about the experiments, so he thought he would start by putting that aside.

"I've heard about the work with RDF, but what other techniques have you tried to find your targets?"

Cunningham smiled.  "Roger was spot on.  You are quite the student."  He took a long swallow of his beer.  "We have tried just about everything a group of demented minds can imagine."  They both laughed.  "We have tried moonlight, starlight, star shell bursts from ack-ack guns, searchlights to bloody awful searchlights on our aircraft, and the only reliable thing we have found so far is plain old luck."  They laughed together again.  "Can you imagine shining the light of a bloody great torch around the sky trying to find a target that does not want to be found?  And, someone who has guns that shoot back rather nicely, thank you."  Cunningham spoke the words with levity in his voice, which Brian responded to in kind.

"So, you feel your way around the sky looking for them?"

"A rather delicate way to put it, actually.  More like a blind fighter pilot."

"But, you've got two night victories already, that says a lot, I think."

"Like I said, Brian, pure dumb luck."

"Somehow, I don't believe you.  From everything I've heard, there are very good reasons they call you, 'Cat's Eyes.'  No offense meant."

"And, none taken.  Let me give you an example of luck."  Cunningham took another long swallow of his beer.  "Several weeks ago, we happened to be on patrol.  Just after midnight, Chain Home reported an inbound single, probably a recce, bandit.  They positioned us using our Pipsqueak and the target return.  They held us in the same patch of sky for the longest time.  There is no telling how many near collisions we may have had, but in an instant my copilot had him against a clear star field.  We nearly ran into the bugger.  I adjusted my position, picked him up myself, and then repositioned for a close range gunshot.  We both lost him several times during the effort, but fortunately reacquired him quickly.  When I had the best firing position I could manage, I squeezed the firing button.  The gun flashes lit up the sky blinding us.  I kept spraying bullets around the point where we thought he was until we managed to hit him.  I then concentrated my fire on those sparks.  Eventually, we got both engines, and he went down somewhere in Oxfordshire, we think.  We fired off nearly a full bag of ammunition at the bastard, and the Blenheim carries quite a bit more ammunition than the Spit or Hurri.  Now, please tell me, was that anything other than blind, dumb luck?"

"I see what you mean."  Both men drank some more beer.  Brian tried to put himself in that cockpit groping around the black night sky purposefully trying to find another aircraft without running into it.  The image was not a particularly attractive one.  "The Blenheim is not particularly easy to see out of, compared to a Spit.  Would it be better to use single seater?"

"We have tried that.  The short answer is, no."

"Why?"

"I think we have given up on the notion of finding them visually, at least until we have them in our gun sights."

"That's where RDF comes in?"

"We are not supposed to talk about that."

"I understand."

"Roger tells me you have done your school work on RDF."

"I was able to visit the Drone Hill Chain Home station in Scotland."

"Smashing.  Well, then, I surmise you must know more about RDF than most pilots."

"I don't know about that."

"I would guess you can imagine the principles being applied to an airborne kit."

"Actually, I'm not sure I can.  What I saw at Drone Hill was enormous.  I can't imagine shrinking that equipment to something that could be carried by a Whitney or Wellington, let alone a Blenheim."  They laughed at the picture of some grotesque, gargantuan appendages attached to a large, four-engine, Whitney Mark V heavy bomber.

"Let it suffice, then, we are working on such equipment."

"Wow."  Brian's mind ground through the possibilities.  If the Chain Home RDF units could find single targets sometimes 100 miles away, maybe the smaller equipment could find them at a closer but still acceptable range.  If Chain Home could get the night interceptors close enough, the airborne RDF could take them in the remainder of the distance for a gunshot.  "How soon?"  Brian asked, using the unsaid connection between pilots.

"A week, a month, a year.  It is difficult to say, actually.  We need it today, or rather tonight, but what we have tried so far does not quite do the trick."

"We all hear the German reconnaissance aircraft more often these days."

"It is only a matter of time before they change out the intelligence kit for bombs. Around the clock bombing is close at hand, I'm afraid. The best we can do is try to take the day away from them, and work to make the night more difficult."

Brian took another few swallows of beer as he considered Cunningham's words. The thought of German bombers filling the night sky as they had most recently done during the day did not make him very comfortable. It was more the notion that he might have to sit on the ground while others like John Cunningham tried to take away the night. Brian remembered the air raid sirens in London, and as was often the case, his nightmare image returned to him. He thought about the possibilities of flying night fighters. While Brian considered alternatives and options, Flight Lieutenant John 'Cat's Eyes' Cunningham excused himself thanking Brian for the stimulating conversation. It was Brian who was most thankful. With two night victories, Cunningham was building quite a reputation as a night fighter pilot although he remained humble and self-effacing. Brian thanked Roger Beamish for the introduction. The night's meager distraction continued until near midnight. Although none of the pilots mentioned the recent lull, they all knew the weather would not stay and heavy German operations would soon return.

———

*Friday, 23.August.1940*
*No.10 Downing Street*
*Whitehall, London, England*

The afternoon, combined, War Cabinet and Defense Committee meetings seemed to drone on, agonizingly slow. The recent respite for the fighter pilots and ground crews had not diminished the strategic and tactical work of the British leadership. While the praise and congratulations of colleagues, advisors, supporters and even detractors from Tuesday's speech was short lived, Prime Minister Churchill did not bask in its glow. Winston found little solace in the Herculean achievements of Fighter Command when viewed against the larger backdrop of the mounting successes of U-boat operations, the specter of night bombing, the lack of adequate countermeasures, the dominance of German armor forces, and the absence of any substantial allied support.

The preponderance of the afternoon meeting dealt with the logistics of sustaining British air and naval operations. The consumption of resources and the tenuousness of their supply lines commanded substantial attention, but Churchill felt a different, more immediate burden. Without air superiority, the invasion would surely come, and the Home Guard and remnants of the

British Expeditionary Force needed time to refit, refresh and retrain. They all knew there would not be much to stop German armor and infantry forces from rolling across England except sheer will power, determination and courage. They simply had to keep the Germans off English soil. If they did not win in the air this summer, there might not be another chance.

"Before we turn to the air situation," Winston said, needing to scratch a particularly annoying itch, "each of you are aware of the enemy's artillery bombardment of Dover, yesterday. We have seen these guns in our aerial photography, but it did come as somewhat of a surprise when they turned those guns to cross-Channel bombardment. When the news came in, I asked the Admiralty to respond. Sir Dudley?"

First Sea Lord Admiral Sir Dudley Pound shifted his position. "As the Prime Minister stated, the recent artillery bombardment from Cape Gris Nez deserves an appropriate response. We have examined the counter battery capable warships available to respond. In addition, we must balance the assets needed to remain ready for opposition of any invasion attempt. We have begun modification of an old dreadnought to add substantial armor above the waterline and focus on the main battery."

A peculiar discussion ensued. The thought of a stationary, floating, large-bore, gun platform anchored in Dover harbor did not seem like a worthy response. The surviving Army artillery did not have sufficient range. The capital ships possessing substantial range would have to take too many risks to successfully engage the German artillery from the confining spaces of the English Channel. As Sir Dudley patiently addressed each question or concern using the logic reviewed many times within the Admiralty, the War Cabinet eventually arrived at the same conclusion. They would proceed with the plans to make HMS *Erebus* operational.

They reached the point on the agenda prepared by Sir Edward Bridges that attracted Churchill's attention. "Sir Archibald, the status of the Air Force, if you please," said Bridges.

The Secretary of State for Air moved his notebook aside, leafed through his papers to find the correct piece of page, and then cleared his throat. "As you are all well aware, the weather has given us a break in the enemy air operations, a much needed break. Before I address several areas, it would be appropriate for Lord Beaverbrook to give us an update on aircraft production."

Minister of Aircraft Production William Maxwell 'Max' Aitken, Bart, Kt, PC, ONB, 1ˢᵗ Baron Beaverbrook of Beaverbrook in the Province of New Brunswick in the Dominion of Canada and of Cherkley in the County of Surrey, member of the Conservative Party, a Peer in House of Lords, and popularly known as Lord Beaverbrook, leaned forward resting his elbows on the table. "We have made substantial progress in the last fortnight. Castle Bromwich has begun delivery of Spitfires while we have managed to increase the production rate at Woolston. Brooklands has stood up to the mark on Hurricane production. Fortunately, German intelligence has not yet achieved the efficiency of their armed forces. None of the primary aircraft production facilities have been seriously attacked."

"That shant last much longer, I dare say," interjected Sinclair.

"Archie is quite right, you know. It is only a matter of time, I should think."

"Lord Beaverbrook has been extraordinarily successful bringing Castle Bromwich on line as well as increasing production at all other aircraft factories. He should be congratulated."

"Here, here," said several of the ministers.

"We have a long way to go," added Lord Beaverbrook.

"Indeed," the Prime Minister said.

"More troubling," Sir Archibald began, "I'm afraid is the precipitous loss of pilots. Sir Cyril and Sir Hugh have taken every step possible to reduce the loss rate. One One Group in the southeast has been the hardest hit. Sir Hugh has shifted assets to the southeast as far as can be safely done without seriously weakening other regions. Pilot fatigue is alarming. We have tried to rotate squadrons to give the pilots a rest, but there are not many fresh squadrons or pilots remaining these days."

"How long do we have at the present loss rate?" asked Churchill.

Sir Archibald looked at his notes. "Two weeks, maybe four if we are lucky," Sinclair answered grimly.

"Is there anything we can do?"

"I am afraid, Prime Minister, that we are doing everything that can be done. We have cut the training course for fighter pilots as short as we safely can. We have reassigned and redirected as many pilots as possible to Fighter Command. Bomber Command as well as the Fleet Air Arm of the Royal Navy have been most generous. Unfortunately, as you may surmise, the new, single seat, monoplane fighters demand a certain skill and *savoir-faire* to fly successfully. The loss rate among new fighter pilots is shocking. To throw more young men into that dreadful environment will not solve the problem."

"What will solve the problem?" asked Clement Attlee.

"Prayers." There was a stir in the War Cabinet. Sinclair did not wait for the questions. "We are in a mortal struggle. We have what we have. All the estimates indicate the window for invasion remains open for another month. We must pray that our lads can hold them off. For the next month, there is insufficient time to retrain existing pilots from Bomber Command, the Navy or any other source. The meat grinder of modern aerial combat, the overwhelming odds our lads are facing every day, and the instant, flashes of these engagements are staggering the imagination, and yet our young men continue to face the challenge unbowed and with unwavering courage, persistence and commitment. We must pray for the good Lord to sustain them in this hour of trial."

The large Cabinet room fell stone silent. Not the slightest sound could be heard. Tears filled Winston's eyes. While intended to be a status report of the Air Force and an answer to a simple question, the Secretary of State for Air offered not just an answer or a report but a prayer for deliverance. Winston Churchill remembered vividly the struggle he witnessed a week ago – the incredible odds; the calm, focused courage of the pilots as they threw themselves into the battle; the precarious edge of the abyss upon which they walked; and, the vigor and determination of the ground crews who braved the explosions of enemy bombs and the guns of diving fighters to refuel and rearm our fighters in unheard of swiftness, skill and their own fearless commitment. Sir Archibald Sinclair had stated their situation precisely and as eloquently as humanly possible.

Winston Churchill knew what had to be done. "We must take the long view, gentlemen. Fighter Command and our young, dauntless pilots, like their predecessors, will stand up to the test, of that I am certain. We must ensure we have left no stone unturned in our efforts to find more qualified pilots to fill the cockpits of Lord Beaverbrook's fighters. Most importantly, our task now is to bolster morale. We must not allow the slightest doubt to germinate in the mind of a single airman, sailor or soldier, and we must encourage the support of our citizens." Winston Churchill scanned the room of attentive faces. He saw the will and commitment in each pair of eyes. "The enemy has focused his attention on Fighter Command and our air defense system. Clearly, their intent is to eliminate our fighter forces, gain at least local air superiority over their planned landing sites, and maintain whatever dominance in the air they can produce. What are their likely next steps?" Chief of the Imperial General Staff General Sir John Dill sat up straighter and leaned forward. "General Dill."

"Placing myself in their shoes, Prime Minister, I should think they will concentrate their bombardments on their planned beachhead areas. With the build-up of motorized barges across the Channel, I suspect it would be safe to assume they are looking at Kent or Sussex, which means One One Group and the southeast air defense facilities can expect even greater attention."

Winston did not wait for other comments.  He wanted to pursue a specific line of reasoning.  "What can we do to deflect them from their focus?"

Winston saw the minds churning away.  Several ministers appeared to have ideas, but probably considered them too extreme or farfetched.  He allowed the thought process to continue several minutes before he made a move to stimulate the discussion.  "Sir Cyril, surely you must have an idea or two."

Chief of the Air Staff Air Chief Marshal Sir Cyril Newall looked down at his papers for a few moments before connecting with the Prime Minister's eyes.  "I share Sir John's assessment, and I have struggled with your question for many days now.  Sir Archibald's portrayal of the air defense situation was precisely correct.  The best we can hope to do is deflect them from their mission."  He hesitated and looked down at his papers once more, as if answers might be on the pages he did not appear to be reading.  "If the Prime Minister and the War Cabinet will permit me some latitude," he waited for a nod from Churchill and Attlee, "we should tweak their nose, their renown German pride.  We should bomb Berlin."

"Dear God, man," said Arthur Greenwood, "they will surely retaliate by bombing London."

"Precisely."

"Are you mad?"

Winston held up his hand.  "Let him continue," he said, nodding to Sir Cyril.

"We must antagonize them to alter their plans.  As you have correctly heard, we only have a few weeks of fighter strength left.  We have no choice but to offer them a more attractive target to satisfy more powerful urges we help to create.  I must say my suggestion is not without substantial risk, but we simply must gain some relief for Fighter Command."

The thought of provoking the Germans to bomb London and other British cities stretched the commitment and enthusiasm of everyone in the Cabinet Room.  The ensuing argument turned raucous for a time as they agonized over the military wisdom and moral consequence of the action.  The reaction, in Winston's mind, came from the surprise of such a suggestion to a consciousness that had not at least asked the question.  He had arrived at the same conclusion several days earlier but struggled with his own conscience.  The fervor of protest began to ebb.

"Then, we are reluctantly agreed," Winston said, as he searched the faces.  Signs of disgust remained but no objections.  "Sir Edward, you should so record this moment.  We shall ask the Air Ministry to instruct Bomber Command to prepare the appropriate plans for a large-scale air raid on Berlin.

I will suggest we hold the plans until the correct moment.  It would be most advantageous if the Germans made the first move to bomb civil targets, but if we cannot find relief, we will take the action unilaterally.  When can you be ready?" Winston asked Sir Cyril Newall.

"Within a day, two at the most."

"Then, so be it."  He nodded to Sir Edward Bridges who recorded the decision.  "It may seem rather *gauche* at this juncture, but how far along are we on the Cabinet War Rooms bunker works?"

Sir Edward answered.  "The major construction is complete.  We are making a few adjustments in the communications system.  We should be fully operational by the 27th of this month."

"We will probably need it soon," mumbled Greenwood.

Winston ignored the comment.  "We should be prepared to move the necessary maps, documents and other items to the appropriate bunker rooms as soon as possible.  Sir Edward has provided the room assignments. We must not disrupt the process of government with this protective move." Winston waited for any discussion.  There was none.  "Now, I would like to hear about our endeavors to improve our night intercept capability."

Sir Cyril received a simple nod from the Minister for Air.  "As the Cabinet knows, we have been working on a variety of approaches.  First and foremost, we are refining our countermeasures against the enemy's *Knickebein* electronic night bombing system and other Nazi bombing aids.  Our night intercept efforts have concentrated on the terminal guidance problem.  Chain Home can position the interceptor fairly well, but the pilot has no help for the final engagement positioning.  We have tested several miniature Radio Direction Finding experimental devices with rather good results.  We expect to have a new Beaufighter configured with the airborne RDF kit by early fall."

"Is there anything we can do to provide an operational capability sooner?" asked Winston.

"The aircraft Sir Cyril refers to will be a hand-built experimental system as it is," answered Sir Archibald Sinclair.  "An extraordinary effort by the Royal Aeronautical Establishment has moved us this far."

"So, the answer is, no?"

"Yes, sir.  Correct."

"We should have our first aircraft through the test program and operational by the fall."

"Lord Beaverbrook has given us considerable assistance with production of the Beaufighter," added Sinclair.  "Our first operational squadron, Six

Oh Four Squadron at RAF Middle Wallop, should be converted and trained by the end of the year."

"Excellent," Winston answered. "I suspect as the days become shorter we shall confront the faceless scourge of invisible bombers in the black of night.  This so called broken leg radio aid system is a prime indicator.  They want to bomb us around the clock in any weather.  The night intercept capability should receive the highest priority."

"Nearly everything must receive the highest priority," protested Sinclair.

"And, we are at war for our very survival," said Winston.  "Resources cannot be a limitation."

"But, they are," said Attlee.

"We cannot allow that.  Could we ever forgive ourselves if we decide to spend just enough, or do just enough, and it proves to be not enough?  We can never predict which idea will work and which ones will not.  We must try every idea, every notion, every thought.  We cannot afford to leave any stone unturned in this mortal struggle.  I must encourage each of you to search out every idea your departments can produce.  A young employee in the Ministry of Supply might possess the key to victory.  Can we afford not to listen?"  Winston surveyed the War Cabinet and did not detect any objections.

"Winston, if I may?" asked Greenwood.

"Yes, Arthur?"

"I heard a rumor this afternoon that Leon Trotsky was assassinated last Tuesday."  Greenwood looked to Lord Halifax.  "What do we know?"

"Our embassy in Mexico City reported the old Russian revolutionary Leon Trotsky, actually born Lev Davidovich Bronshtein, was assassinated at his home in exile in Coyoacán, Mexico City, Mexico.  The details are still being collected quietly, to avoid drawing attention to the crime or our interest in the crime.  Trotsky was struck in the head with a mountaineer's ice ax.  The blow did not kill him outright.  He died the next day.  We strongly suspect Stalin sent one of his most trusted NKVD agents to eliminate the last of the old guard and his only potential challenger.  The police investigation continues.  Our folks in Mexico believe the assassin successfully escaped and may never be identified."

"Stalin is a ruthless thug, with no conscience, and is only interested in absolute, dictatorial power."

"We have no indication Trotsky had any interest in returning to Russia or engaging any of his old political cronies.  Trotsky was not a threat to Stalin's power, but he murdered him anyway."

"We have been at this for nearly four hours. I, for one, need a modicum of respite. If there are no objections, we shall adjourn for the evening meal and reconvene at 21:00 hours sharp." Again, he scanned the group and saw no objections. "Done."

———

*Saturday, 24.August.1940*
*RAF Middle Wallop*
*Middle Wallop, Hampshire, England*

"This is perfect flying weather," said Brian. "Where are the Germans?"

"Easy, laddie," answered Brian's section leader, Flight Lieutenant Roger 'Jackstay' Beamish. "They'll come soon enough."

"It's this damn waiting that is the worst."

"Maybe for you," added Pilot Officer Stanley 'Slim' Koenig, one of the other American volunteers in No.609 Squadron. "Some of the rest of us would rather they not come back."

"I know, but we all know they are coming back, and sitting around waiting for the birds to come to you is not the best way to hunt pheasant."

Brian's country humor struck a chord with the group, sparking off a flurry of light-hearted jabs and pokes at one another and their adversaries. None of them liked the waiting no matter how they presented the situation. Some feared the combat more than the fatigue and tension of the wait. Brian felt more in control amid the chaos of aerial combat than he did letting some unseen adversary dictate the time of their meeting. He would rather go out and flush out the Germans rather than wait for them to come to him.

A faded, green, Land Rover drove down from the Operations building and deposited Squadron Leader Darling. Wing Commander Worly, the station operations officer, released the steering wheel and joined Darling. The leader of No.609 Squadron motioned for the attention of the pilots. He waited until the few napping pilots were with them.

"It seems, gentlemen, we have an award ceremony to go to next Thursday given we are not in the air." The pilots looked around, all of them wondering what the announcement meant. "As Wing Commander Worly, here, received the message, Flight Lieutenant Beamish and Pilot Officer Drummond are to receive the Distinguished Flying Cross as *bona fide* aces." Congratulatory back slaps and handshakes passed between them. Darling and Worly shook hands with both men, and then Darling cleared his throat to regain everyone's attention. "What will make this award ceremony more special is, the King is to award the George Cross to Mrs. Ian Palmer, the woman who saved our young colonial ace's life several weeks ago."

Hoots and hollers followed.

"My Christ," interjected 'Waggle' Davies. "Isn't the George Cross the newly created, highest civilian award for courage?"

"Yes, it is," responded Worly.

"I mean, we love you dearly," Davies said, looking and nodding to Brian, "but, all the woman did was pull him from the puddle."

"Risking her life with the wet and sinking parachute. Both 'Jackstay' and 'Harness' witnessed her struggle."

"Maybe she just could not swim," added 'Boxer' Stockard, instigating more gibes.

"Apparently . . . gentlemen, please . . . ," Darling tried to continue, "apparently, His Majesty's Government wants to make this presentation a very public event to bolster morale. Mrs. Palmer lost her husband aboard *Glorious* this last spring and risked her life to save a very young American volunteer fighter pilot who landed, unconscious, in her pond after shooting down two German aeroplanes." Ten 'VK' Hurricanes from No.238 Squadron took off passing over the tree tops just in front of them making Worly visibly nervous. He was not at his post, and the action was beginning. "We don't have much time, it appears. Business may be picking up. At any rate, the British media as well as much of the world press who remain in the country have been invited. They want the world and especially our American cousins to see what our own Mister Drummond has been up to in this little fracas."

Wing Commander Worly shook hands with 'Jackstay' Beamish, and then extended his hand to 'Hunter' Drummond. "Congratulations to both of you. We shall look forward to your recognition next Thursday. Now, I am afraid I must run, as Squadron Leader Darling said, business appears to be picking up. Good hunting, gentlemen," he said and then drove back to the tower.

The squadron did not enjoy much time to savor the moment. The telephone rang before Wing Commander Worly reached the Operations Tower. "Scramble the squadron," shouted Corporal Warren from inside the dispersal tent. The pilots grabbed their flying kit and ran to their aircraft. They were airborne within minutes.

Brian took one brief moment as the squadron climbed to their intercept altitude to think of the award recognition for what he was not precisely sure, but more significantly he would see Charlotte Palmer again in a more public, neutral environment.  Maybe she would not feel threatened.  Maybe she would see him in a different light that might enable a more lasting relationship.  His moment did not last long.

The Hurricanes of No.238 Squadron missed the raid on Portsmouth, as virtually every squadron had experienced many times.  Their sister squadron returned to RAF Middle Wallop low on fuel.  By the time No.609 Squadron reached Portsmouth, at least one large formation of Ju88 medium bombers were nearly complete in their bombing run on the shipping and port facilities.  Several ships were burning as well as a number of buildings near the docks.

Squadron Leader Darling led the pilots in an attack dive toward the turning bombers.  Before they reached the bombers, 'Fog' Johnson called the enemy fighters falling upon them to protect their bombers.  They all pulled up turning into the hive of diving Bf109 fighters.

The ensuing air battle mimicked so many others.  While there was order at the beginning, the engagement almost instantly degenerated into a free for all of twisting, diving, turning and climbing fighters mixed up in a great ball.  Their opponents this day were red nosed, Bf109E-4 fighters of JG52, and they were very good.  They made no mistakes, and fortunately neither did the pilots of No.609 Squadron flying their Vickers-Supermarine Spitfire Mark IA fighters.  Numerous hits were achieved on both sides although none of them were apparently serious.  'Slim' Koenig got the worst of it.  He reported his coolant temperature climbing, which meant he probably had a hit in the ethylene glycol coolant system.  He diverted to RAF Warmwell.  'Waggle' Davies and 'Fog' Johnson disengaged to make sure he got there safely and did not get jumped by any rogue Germans as they descended.

As the Germans disengaged, the British gave chase for a short time, and then broke off themselves several miles out over the Channel.  Although No.11 Group Instruction No.4 did not apply directly to the pilots of No.10 Group, they all knew the reasons behind Air Vice-Marshal Park's order, and those reasons applied to them.  The ships and dockyards of Portsmouth were burning black and ugly as they turned northeast and joined up on their leader.  The squadron, less Koenig, returned to RAF Middle Wallop.  Several of the aircraft had bullet damage, but none of them displayed any signs of large damage usually done by the explosive shells of the German 20mm cannons.  They arrived too late to stop the bombers and fought to a draw with a superior number of fighters.

They gathered as they always did after their mission debriefings outside their dispersal tent at the southwest corner of the fighter station.  They settled into their chairs or lay on the grass.  They waited for the next challenge.

"Did 'Slim' get down safely?" asked 'Harness' Kensington.

"His airscrew was still turning when he touched down," answered 'Fog' Johnson.

"Took a coolant hit?"

"So he said." Johnson's response was the closest any of them ever went to acknowledge Koenig's lack of enthusiasm for the fight. None of them truly enjoyed the danger, but they usually threw themselves into it with vigor. Koenig was the exception, and they understood.

"Word from the east is, Manston took a severe pounding," announced 'Boxer' Stockard. "Closed the place, the landing area was so chewed up. Biggin Hill, Croydon and Kenley got hit pretty hard as well."

Corporal Warren joined the discussion in a rare moment. "My girl-friend said they evacuated Manston because things were so bad."

"That's a hellava deal," Burns said.

"Any of our stations get hit?" asked Davies.

"According to Group, none. Just the port facilities at Portsmouth, of which we are fully aware," answered Darling. "They appear to be concentrating on One One Group and the southeast."

"They're getting serious about this invasion shit, aren't they?" added 'Red' Burns.

Several of the pilots laughed to relieve some inner tension. The signs were all around them. They all felt a certain amount of impotence to stop the inevitable invasion. There were no more questions of whether; now, it was only when.

"You might say that," said Davies.

"The other piece of significant news from Group is, they apparently bombed London this afternoon. Several other cities were bombed as well."

"That ought to make Mister Churchill angry," 'Sparky' Morrow said.

The attacks on the southeast continued through the afternoon. No.609 Squadron was brought to Standby several times in anticipation of the need for support of No.11 Group squadrons. They received reports throughout the afternoon of the devastating attacks on the southeast. Their sister squadrons were being chewed up pretty good. Numerous aircraft were being lost on the ground, as they tried to refuel and rearm. Brian and his colleagues wanted to help, but the orders did not come. They stayed on the ground, listened to the sketchy radio reports, and tried to catch snippets of information from friends in the east. 'Slim' Koenig returned with his 'PR-R' Spitfire just before the squadron was released for the day. The evening was spent trying to gather up information from their friends, compare notes and create an image of the day's action and what had changed.

Brian thought of Jeremy Morrison who had to be in the thick of the action. He also thought of Mary Spencer, Rosemary Kensington and Charlotte Palmer. Brian considered several times calling Charlotte to make sure she was all right given the day's action, but he did not have her telephone number. Brian knew she had a telephone, but he never asked for her number. He would have to wait to hear about her.

—

# Chapter 8

Concentration is the secret of strength.

-- Ralph Waldo Emerson

*Sunday, 25.August.1940*
*RAF Warmwell*
*Warmwell, Dorset, England*
*08:45 hours*

### Week 8

The long, broad nose of the Vickers-Supermarine Spitfire Mark IA along with the image of the 1,030 horsepower Merlin III engine purring at low cruise power under the cowling always gave Pilot Officer Brian Drummond a secure feeling.  He possessed a unique confidence in the machine that fit him like the ultimate, soft, supple, tailored leather glove.  As he surveyed his machine, he could find no holes, jagged edges or other discontinuities.  The only signs of recent events and the successful mission were the tattered, blackened and flapping remnants of the red gunport tapes, and the black smudged streaks produced by muzzle gases streaming back from the eight gun ports.  Brian's 'PR-F' Spitfire flew proudly with a cocky confidence that connected the machine to him in every way.  The small confines of the Spitfire cockpit enhanced the sensations of intimacy for Brian.  He could depend on this living, breathing, biting, flying machine to respond to his every command with precision and enthusiasm.  He had never felt so comfortable in the cockpit of any other aircraft.

Brian scanned the other airplanes around him.  Normally, he was only concerned about shapes and the alignment of formation checkpoints to ensure he maintained the proper position relative to his section leader.  This time he looked around to see the same signs of combat on the other aircraft.  He saw holes, torn aluminum skin and jagged pieces flailing in the slipstream on several of the neighboring machines.  It had been a hard, but successful fight.

"Warmwell Tower, this is Sorbo Leader calling with a flight of twelve for landing," Squadron Leader 'Spike' Darling radioed, as they approached RAF Warmwell.

"Sorbo Leader, this is Warmwell.  The aerodrome is temporarily closed for repairs.  Afraid you will have to divert to Middle Wallop.  Alighting area unusable.  We were hit by Gerry about an hour ago, just after you took off."

"Why didn't Bandy tell us?"

"Don't know.  We expect to be operational in another few hours."

"Roger, Warmwell.  Sorry we were tied up and not here to help."

"Understood, Sorbo Leader.  Thanks for the thought."

Darling turned the flight northeast toward Middle Wallop, leveled off at 1,500 feet, and throttled back to save fuel, cool off the engines, and reduce wear and tear. Their home base had not been attacked, so far this day. They landed without incident. The ground crews swarmed over the Spitfires with the sole objective of making them ready for launch as quickly as possible. There was poetry to the sea of motion around, under and over the sleek aircraft that could be appreciated by anyone who observed. These airmen were good, very good, at what they did, and the pilots respected their expertise. The flyboys could not do what they needed to do without the ground crews and their skills. It was the most tightly integrated symbiosis Brian had ever experienced, and it never ceased to amaze and wonder him.

"Everything tip top, sir?" asked Leading Aircraftman Bernard Gordon, as he helped his pilot extricate himself from the snug cockpit.

"Like the beauty she is, Bernie," he said, as he stepped out onto the left wing and patted the back of his crew chief's left shoulder. Brian jumped down off the wing and stretched his muscles, as he did every time he returned from a long mission. He looked under the wing to see Aircraftman Colin Jenkins, his armorer, and Aircraftman Jordan Toldson, his mechanic, working on the guns. The access panels for all four left wing machine guns were open allowing Jenkins to check each breechblock and feed mechanism. "They were perfect, Colin. Keep up the great work."

"Aye, sir. Any more victories with me ladies?"

"Not this time, but we got a bunch of hits."

"Smashin', sir. We'll keep me lit'le ladies workin' for ya."

Only half the debriefings had been completed when the telephone rang. They continued with their activities waiting for the command to launch.

"Gather up, lads," shouted Darling. He waited until all the pilots were attentive. "Sector reports a large raid headed toward Weymouth, again. Since this raid appears to be the only one in our area, they are going to launch several squadrons. When our birds are readied, we shall take to the air and join the group. Although only half the aeroplanes have been rearmed, I would suggest we mount up. As soon as the last bloke is ready, we shall start and have another go at Gerry."

Without words or any recognizable acknowledgment, the pilots walked silently toward their respective fighters. Jenkins, Toldson and Gordon were finishing the loading and priming of the last gun on the right wing as Brian approached. Gordon checked the others and left them to finish the task.

"She'll be ready in a snip, Mister Drummond," Gordon said.

"Thanks, Bernie."

"They barely gave you time to catch your breath this time."

"The way it is, I'm afraid. Gerry isn't giving us much of a break."

"Must be thinkin' they'll come join us over here," he laughed.

"Not much chance of that as long as these guns keep working," Brian answered, waving his leather helmet, goggles and oxygen mask at the wing.

"Then, we shant be havin' schnitzel and kraut for lunch anytime soon, ya say."

Brian laughed, causing the others around the 'PR-F' Spitfire to laugh as well. "Not if we have anything to say about it."

"Excellent."

"Bloody good," added Jenkins, as he came around the tail to join them. His dark blue coveralls were heavily stained with black gun soot and grease. "Me ladies are ready for ya, sir."

"Thanks, guys."

"Good hunting, sir," said Gordon.

"Here, here," added Jenkins and Toldson.

The string of twelve Spitfires snaked their way up the rise along the western edge of the landing area. As they taxied, the No.238 Squadron 'VK' Hurricanes took off *en masse* toward the South. Brian glanced over his shoulder to see the ten fighters bank right. They were probably headed toward Weymouth as well. No.609 Squadron followed them shortly thereafter.

As the pilots of No.609 Squadron climbed and prepared their fighters for combat, Middle Wallop Sector Control provided their intercept vector instructions. The raid was reported to be in excess of 200 enemy aircraft. As best Brian could ascertain from the radio traffic, the majority of six Fighter Command squadrons, approximately half Hurricanes and half Spitfires, probably near 60 or 70 fighters, were climbing to meet the raid. The developing clouds obscured most of the horizon up to 10,000 feet. As they broke out above the tops, the sky was clear above. Several more minutes passed before they found their target.

The squadron was passing through 16,000 feet climbing to 24,000 feet when the German raid became clearly visible. The scene possessed a strange fascination. The bombers were concentrated in one large group with at least two types of medium size bomber aircraft. What set this raid apart from the others Brian had seen was the number and distribution of escort fighters. There appeared to be at least twice as many fighters – all Bf109s. This was a serious raid. The escort was split into three distinct sections. Half of the German fighters were split into two formations on either side of the bombers at roughly the same altitude. The other half of the escort fighters hovered 5,000 to 7,000 feet above the bombers. As his altimeter continued to wind up, he saw the converging RAF fighter squadrons.  One Spitfire and three Hurricane squadrons were closing rapidly on the bombers and fighters stacked up between 12,000 and 15,000 feet.  The other Spitfire squadron approaching from the west had to be from Pembrey in South Wales.

The formations converged.  As Brian watched the initial moves of the lower aircraft, he kept an eye on the higher group of Bf109 fighters who waited for the closing British and rechecked yet one more time the readiness of his machine for combat.  In another minute or two, the seemingly slow motions of the aircraft would turn violent, purposeful and dangerous.

As the groups of fighters at the higher altitude closed with one another, the Germans split into two response sections climbing to deny the altitude advantage to their opponents.  "Here we go now, lads," was all 'Spike' Darling radioed, as he turned the squadron for the final time and headed directly into the German fighters.  There were twice as many Bf109s facing the British fighters.

Once the first shots were fired, the sky narrowed quickly, and Brian's head and eyes moved quickly, precisely and skillfully.  Targets to shoot at were everywhere as well as compatriots.  The secret rested in the hunt for the hapless adversary too intent on his own target, or marveling in the gloriousness of his moment of victory.  Brian found one, then another, and dispatched them with surgical efficiency.  He also helped several others to shake their attackers as he received helped a few times.  Awareness and the ability to adsorb and respond to what was going on around him was the key to his survival and success.  Altitudes continued to creep up as each fighter pilot, friend or foe, tried to find that slight margin of advantage.

The large engagement above the approaches to Weymouth and Portland Harbor lasted less than an hour.  A very small number of the German raiders reached their targets.  The sea was pocked with spots of burning fuel, the distinctive circles of exploded bombs and the telltale signs of battle.  The German fighters broke and disengaged en masse with their fuel running very low and the bombers depositing their lethal loads mostly in the English Channel – all in all, a successful mission for Fighter Command.

———

*Sunday, 25.August.1940*
*Cabinet Room*
*No.10 Downing Street*
*Whitehall, London, England*
*16:00 hours*

The days blended together even for the leadership of Great Britain during that fateful summer.  The Defense Committee gathered as they did nearly every day now to assess events and make decisions regarding the defense of the realm and the empire. These were not easy days.

In addition to the Minister of Defense, the full committee attended – all three defense ministers, all three chiefs of staff of the military departments, all three military members of the Cabinet Secretariat staff as well as General Ismay, and on this afternoon, Patrick Francis Kinna, Churchill's principal stenographer, would be the only note taker for the meeting, since he was the only one in the stenographer pool with sufficient security clearance for the day's topics. Also in attendance for this afternoon's meeting was Home Secretary Sir John Anderson.

"Shall we begin," announced Prime Minister Churchill. The room quieted immediately. "First and foremost these days, let us have the latest situation report on the air campaign."

Air Minister Sir Archie Sinclair began, "Yesterday was a dreadful day, I'm afraid. The enemy mounted rather large raids all across the southern sectors. As has been so common of late, One One Group paid the heaviest price. We lost two-dozen aircraft and nearly that many pilots out of the southeast group alone. Several aerodromes remain unusable as the Germans bomb the repair crews before they complete their work. They have tried working through the night, but they are making little headway without lights."

"Can we redeploy troops? Would more people help?"

"No, I'm afraid not. There are limits to how many hands can be in the bread dough."

"What about the pilot losses?"

"I am reluctant to note here, our situation is more dire." Sinclair paused to gather his thoughts. "If you recall Dowding's projection chart from his briefing three months ago, our net loss rate has accelerated. Sir Hugh has already ordered rest periods cancelled, reassignments postponed, and training cycles shortened to the absolute minimum. Sir Cyril has ordered all training command pilot output temporarily diverted to Fighter Command. We are making every effort to maintain fighter strength. Lord Beaverbrook's masterful aircraft production achievement has at least kept aircraft net loss ahead of the pilot lossess."

"How much time do we have?" asked Churchill.

"At the current rate, we have a week . . . two at the outside."

"What happens at that threshold?"

"Fighter Command will be unable to mount an effective defense of our air space. We will have to make critical decisions regarding limited defense of specified locations. We will not be able to defend them all."

"Some will argue we have not been able to do that from the beginning of this fracas.  As Stanley Baldwin said all those years ago, the bomber will always get through."

"Yes, well, we cannot defend everything all the time."

"Some will argue we cannot defend anything some of the time."

Sir Archie uncharacteristically lost his temper.  "What would you propose, Prime Minister?"

"Easy, Archie.  I am not the enemy.  I am only a simple realist who bears witness to events around us.  I share your sensitivity.  These young, intrepid aviators have done yeoman's work against an enemy with overwhelming numbers.  I do not disparage their effort.  If anything, I condemn those decisions made decades ago in the name of peace that allowed the Germans to build such a fighting force and left us woefully ill prepared to deal with that enemy.  No, Archie, I am not cross with our airmen or the Air Minister, or his ministry.  Yet, here we are.  This is our mess to deal with, no one else's.  So, you proposed to gradually draw in our air force to conserve our dwindling resources to presumably protect the capital."

"Actually, yes and no.  Yes, we will soon need to contract our forces to maintain mass, and no, I am not suggesting we protect the capital, but rather our aircraft production capacity.  Pilots are presently our most critical resource, but without aeroplanes for them to fly, they are just skilled craftsmen with no tools."

"Dear Lord," mumbled someone.

"What about Leigh-Mallory's Big Wing tactics?"

Chief of the Air Staff Newall squirmed in his chair.  Sinclair chuckled softly, more to himself.  "Simply put, Prime Minister, there simply is no time, unless we are prepared to abandon the aerodromes of Kent and Sussex.  And, if we do that, we are conceding the southeast to the Germans and their invasion plans.  If we are going to perish, I say we go down fighting to keep Gerry off our beaches.  The *Bosche* have paid a hell of a price for what they have achieved.  But, they have not come, yet."

"They are still training and preparing," interjected War Minister Anthony Eden.

"Yes, I am certain they are, but one of their essential requisites – local air superiority – has not been achieved."

"Yes, yes," Churchill said impatiently, "but, Archie, the question is when will they attain that threshold and pull the trigger on their damnable invasion."

"My guess is, as I tried to say earlier, one to two weeks at most.  Sir Cyril, do you have anything to add?"

"No, minister.  The Air Staff concurs with that assessment."

"Well, then, the time has come to change the equation," Churchill pronounced. "As we discussed several days ago, we must distract them from their objective. If the reports I received this morning are correct, German bombs fell on the outskirts of London."

"Yes, sir, that is correct," Sir John said.

"Then, we have our justification to bomb Berlin, as we discussed.

"We have already taken the liberty, Prime Minister," responded Sir Archie, "as we had previously agreed, to launch a good size, night raid on the armament factories in the northern sections of Berlin. Sir Cyril, if you would . . ."

"The bombers have been loaded. The crews have been briefed. They are completing their final preparations for the long night navigation to Berlin. They will launch the raid after evening twilight . . . in a few hours, unless we cancel the raid."

"God help us," someone mumbled.

"You know what will happen," Sir John interjected.

"Yes, I do, John. We will smite directly the arrogant pride of that vile, boastful dictator, and he will order the *Luftwaffe* to attack London as public retribution for the insult to his ego."

"Innocent civilians will die, Winston, for God's sake."

"Yes, John, I am afraid so. This is the nature of total war. Londoners will take the beating to keep the Germans off our precious island."

"There is no guarantee."

"No, John, there is not, none, no guarantee this tweak of his nose will work, but as Archie described for all of us, we are rapidly running out of options here."

"But, still . . ."

"But, still, nothing, John," Churchill said more firmly than he intended. "We must stop them before they have a chance to cross our moat. We must all sacrifice to buy our fighter pilots breathing space. Unless anyone has a better idea, we launch Bomber Command on that raid to Berlin tonight." Churchill scanned the room to see negative gestures and silence. "Then, so it shall be. Let us pray those bomber crews hit their targets and Corporal Hitler takes the bait." Churchill bowed his head for a few moments. "With that nasty business begun, I must ask, Sir John, when will the Cabinet War Rooms be ready for occupancy and operation."

"We have taken advantage of the last few weeks to refine the accommodations underground. We can use most of the bunker at any time, although it would be best to let the construction crews finish their current tasks. We should be fully ready early next week."

"None too soon given tonight's decision," added Eden.

"Yes, quite so."

"Very well, then. We must anticipate Hitler will swallow the bait whole and set the hook. Number Ten will undoubtedly be a significant target for the Germans as well as Whitehall and Westminster. Please let us know as soon as the War Rooms are ready." Sir John nodded his agreement. "As soon as Sir John gives us affirmation, we shall move the War Cabinet and Defense Committee meetings to the War Rooms conference room, or if we have signs of aerial attack on London sooner, then we should shelter here so we can maintain continuity of leadership during the attacks. Is there anything else we need to discuss tonight?" Churchill searched each set of eyes. "Very well, then. Hopefully the plan we have set in motion works as expected. Have a good evening, gentlemen." Churchill was the first to depart the Cabinet Room. He did not wait for the others and proceeded to his office.

—

*Sunday, 25.August.1940*
*RAF Middle Wallop*
*Middle Wallop, Hampshire, England*
*22:30 hours*

From the intelligence debriefing, the tally for the afternoon's entanglement was placed at five German bombers down, a score damaged, four Bf109s shot down with two parachutes, against the loss of three Hurricanes and two Spitfires with only one known fatality. The pilots had plenty to discuss at the evening's gathering at the Officer's Mess bar. The evening meal passed quickly with unusual animation for the tired crews. Gradually, everyone moved into the small space of the bar, and then out the back door into the garden patio area to enjoy the summer evening.

Then, someone said, "Do you hear that?" The conversations stopped instantly, as they all listened for the beating drone of the German night bombers. The sound was distinctly different. Brian and Jonathan joined others walking through the blackout curtains to a star filled night and low quarter moon on the eastern horizon. The heavy humm of synchronized airplane engines, many of them overhead, told Brian they were not Germans.

"Our blokes going somewhere," suggested one of the pilots.

"How can you tell?" asked Jonathan.

"Airscrews are synched. Our lads manage to synchronize their airscrews, so you don't hear that god-awful drum beat like Gerry. The bastards never seem to figure out how to synchronize their airscrews."

"Sounds like quite a few."

A dozen assorted pilots stood outside in the dark with their heads craned toward the heavens vainly searching the skies in the direction of the sounds.  The deep, low groan filled the sky above them and felt big.

"If they are our guys, where are they going in the middle of the night?" asked 'Red' Burns.

"I'd say they are going to pay Gerry a visit."

"It seems like they're headed east, not south."

"Germany," someone offered.

"Hellava deal, if that's so," commented Burns.

Brian remained focused on the fading sounds to the east.  The moment allowed him to recall his unasked questions about night bombing.  He remembered his discussion with 'Cat's Eyes' Cunningham as well as his own experiences with night flying in Kansas and Great Britain.  Finding a darkened airfield was a very difficult, if not the most difficult, task Brian had done.  The thought of many bombers finding their targets and precisely dropping their bombs on the desired target overloaded his imagination.  "How do they find their targets?" he asked matter of factly.

"Damned if I know."

"Beats me."

"Magic."

The series of casual comments from unknown speakers produced a string of laughter that broke the mood.  The sounds of the bombers had gone.  The fighter pilots returned to the bar and the joviality contagiously spread within the Mess. The mood was good despite the burdens of combat, maybe even in defiance of the burdens of combat.

"Since they bombed London, do you think they are headed toward Berlin?" asked Jonathan.

"Maybe," was the only answer he received from several of the other pilots as everyone returned to lighter topics.

The energy in the RAF Middle Wallop Officer's Mess bar dissipated early, as it so often did in these long, torturous days of summer.  The intensity of the air battle by day, while it did not dull the humorous, defiant edge of the fighter pilots, took a good portion of their capacity for frivolity and merriment.  Most retired before midnight.  Brian fell asleep with questions about pilotage of night missions of Bomber Command.

———

*Monday, 26.August.1940*
*RAF Middle Wallop*
*Middle Wallop, Hampshire, England*
*16:10 hours*

The overcast kept the temperature at a very comfortable level all morning. The waiting and last night's questions offered fertile ground for discussion, speculation and conversation. The pulsating beat of the occasional German reconnaissance aircraft high above the clouds reminded them of last night's sounds. For Brian, the contrast added to his inquisitiveness. The lack of even the slightest hint of gunfire or change in the monotonous tones probably meant the intruders were unchallenged. Brian wondered whether the heavy, four-engine, Whitney Mark Vs of Bomber Command had been opposed near their targets, wherever they were located. Had the Germans managed to figure out what Fighter Command was still struggling to find – a viable option against night bombers?

Flying Officer James Royster, the squadron's intelligence officer, walked smartly down the edge of the landing area from the Operations building. He appeared to be a man with a purpose.

"Wonder what our resident spook has for us?" asked Flight Lieutenant Robert 'Sparky' Morrow.

"Does appear to have a measure of urgency to his gait, doesn't he?" Flight Lieutenant John 'Waggle' Davies added in a casual, off-handed manner.

They waited for the slight intelligence officer to reach the dispersal tent. Even Squadron Leader Darling and Corporal Warren came out of the tent to join the others. None of them said a word. As Royster approached the tent and the group of faces looked his way, he could not contain his enthusiasm.

"You would not believe what the bloody frigging Bomber Command did last night," he said, as he continued to walk the remaining distance. No one answered him. Each of them gave him space to continue. "The sodding blokes bombed Berlin," he said nearly shouting. Cheers and clapping punctuated the news. "According to the Air Ministry, they hit their targets, caused holy hell to erupt and otherwise shook the bugger's cage."

"About damn time we took it to them," Davies said.

"Taste of their own medicine," added 'Red' Burns.

"I'll bet we see the mark of your chum, Winston, there, 'Hunter,'" Morrow said.

Brian liked the thought, but felt a twinge of embarrassment. He had only met the venerable statesman once. That hardly qualified as chummy. There was also a stream of pride among the other thoughts. Winston Churchill had proved to be everything Malcolm Bainbridge had told him more than two years ago before all this happened.  The British leader made it easier to focus on the task in front of his gun sight.

"So, what do you think, 'Hunter?'" asked 'Jackstay' Beamish.

"I think 'Sparky' is right.  Bombing Berlin is just what Winston would do."

"Smashing," shouted Morrow.  "We were bloody well right.  He is chums with the PM – bloody friggin' Churchill."

"Now, now, 'Sparky,' don't let your Labor roots and Irish sympathies show too much," added Davies.

They all laughed.  The image of British bombers over Berlin had a powerful impact on Brian's commitment.  No one was going to roll over and wait for the end.  The perseverance of good over evil, and eventual success.  This was just as Malcolm had described it.  The connection between these men standing against an enemy.  The purity of their purpose.  The struggle for victory.  This was what Malcolm had always told him.

The telephone rang, and the conversation stopped instantly.  Mansek and Stockard instinctively gathered up their flying gear and took several steps toward the standing fighters.  "Scramble the squadron," shouted Warren from inside the tent.

Darling emerged at the same moment.  "Let's hop to it, lads," he said, as he started to run toward his machine.

They were airborne within two minutes . . . very fast given their current dispersal location.  The twelve machines gathered up close as they entered the cloud deck.  Brian narrowed his focus to only one point – the outboard corner of the flap lined up with the aft most exhaust port.  The alignment of the checkpoint kept him in proper position relative to his section leader, 'Jackstay' Beamish, in the 'PR-D' Spitfire.  They broke through the overcast at about 8,500 feet, but the scattered clouds continued well above 20,000 feet.  Their intercept vector leveled them off at 17,000 feet.

As they approached their intercept point, Squadron Leader Darling spread them out into their attack formation, allowing each of the pilots to scan the sky around them for targets.  Several minutes passed before a formation of 10-15, H-tail, twin-engine, Do17Z-2 medium bombers came into clear view.  Before Darling moved to attack, they all searched the sky above the bombers.  It had been many months since they found unescorted medium bombers.  Brian remembered the Ju87s attacking RAF Middle Wallop unescorted.  The result would soon be the same.  No.609 Squadron had been positioned perfectly for the attack on the vulnerable bombers.

Darling rolled his fighter for the attack dive from the right, rear quarter of the German raiders with the sun directly behind them.  The others in turn followed their leader in the attack.

Brian picked out his target, adjusted his flight path for a perfect rear quarter slashing attack, and took one last look over his shoulders before concentrating on his target.  The weak flash above and behind him told Brian there were probably Bf109s rolling in for their own attack.

"Bandits, seven o'clock high," called Brian over the radio.

He pulled back on his stick and rolled left to bring his nose up to the diving German fighters.  The shapes quickly appeared in front of him.  Something like twice their number caught them at a disadvantage.  Brian and his colleagues were now climbing into the Germans.  The youngest American found his best target, aligned his sight, quickly checked and centered his slip ball, waited until his target approach the correct size in his sight and pushed on his red firing button.

The startling weak pop of only one round going off from some of the guns left Brian cursing the failure as he pulled hard to disrupt the adversaries firing solution.  The two groups of fighters passed each other in a flash as they all turned and climbed.  Brian found the enemy fighters and took the moment to recheck all his switches.  Sharp control inputs brought his nose onto a still turning German.  Brian pushed the firing button.  Nothing happened.  The sickening sensation in his gut did not stop him from rolling his Spitfire inverted and pulling his nose through to deny the German any chance for a shot.

"My guns are jammed," Brian radioed.  "I'm breaking off."

"Me, too," 'Crazy' Kradilcek spat.

A strange anxiety gripped him like a weighted rope entangling his foot and dragging him toward the cliff's edge.  Here he was in the middle of a fresh fight with a full load of ammunition and eight jammed guns.  One or two guns have malfunctioned before, although not often, but never all the guns.  Several others radioed the same experience.  Something dreadful had happened to the guns.

The extrication process took much longer than he wanted.  His best defense until he could find the right opening was aggressive flying.  Their opponents probably did not know there were no bullets coming from some of the British fighters.  The helpless feeling of not having control, not being able to take the fight to the Germans, not determining the outcome as a hunter, contributed to Brian's sickness.  The only thing saving him from another swim or death was his prowess at maneuvering the machine.  His eyes found the threats and helped him deny the Germans a good angle or a good shot.

As his flight path came close enough to a large, billowing cloud to make a try, Brian heard his leader's radio call. "Sorbo, disengage."

The command directed each of the pilots to make their move to dive out of the fight. The cloud was close enough, enabling Brian to dive into it. The combination of his high speed and the natural turbulence within the cumulous cloud shook his airplane violently, causing noticeable pain in his torso as his organs banged into one another. The violence of jolts to the aircraft blurred his vision, adding to the vertigo sucking him into confusion and disorientation. He fought to maintain control of his machine for the longest minutes he had experienced in quite some time, and then he saw open sky, but only for a second.

Brian struggled to keep his wings level, as he slowly tried to bleed off airspeed. As he did, the violence of the turbulence decreased in severity although the deceleration added substantially to his saturating sensations of disorientation. Another brief glimpse of sky did not give him enough time to pull his nose into the hole before he was back in the clouds again. The airplane began to gradually settle down as his airspeed continued to decrease along with his altitude. At about 5,500 feet another hole opened up in the clouds. Brian banked the fighter onto its right wing and pulled his nose into the clearing. It was a large gap in the clouds with no blue sky detectable anywhere. Brian slowed his Spitfire to just above approach speeds as he searched the bubble of open air around him. He saw a short view of another Spitfire on the other side of the bubble passing back into the clouds. Brian completed the remainder of his descent through the clouds until he could see the dark blue-green of the cloud-covered sea, the white ribbon of the beachline as the rich green and brown colors of the trees and fields beyond the beach.

Darling's radio calls to his squadron directed them to rejoin under the clouds near the port city of Poole. Stockard and Burns complained about their gun jams until Darling put a halt to the extraneous radio chatter. The short flight back to RAF Middle Wallop concluded in silence. Brian could only go through in his mind what had happened to his guns. He had a few gun jams and ammunition belt breakages, but he had never had all eight guns jam at once. The scene they just departed – jumping into a fight with first line German fighters, superior in numbers, without working guns – sent shivers through his sweat cooled body.

As they taxied to their parking spots, Brian saw Gordon and the crew running toward the aircraft. When they noticed something was not right with their fighter, all three expressions changed. They slowed to a walk.

"What happened, sir?" asked Bernie Gordon as Brian unstrapped from the Spitfire.

"All the guns jammed on the first round, Bernie," he answered with pronounced irritation in his voice. "Bloody damn guns jammed at t he start of a brawl. That's never happened before. Whatever the hell happened, we had better fix it pronto."

Gordon seemed staggered by the news. The color washed from his face, as the leading aircraftman realized the fear Brian felt in the air. Gordon jumped off the wing as Jenkins emerged from beneath the wing.

"It looks like only one round was fired," Colin Jenkins said with surprise.

"Mister Drummond says they jammed on the first round."

"Bloody hell," Jenkins responded, scratching his head. "The guns look like they are in perfect working order."

"Something is amiss, Colin."

"Nothing obvious."

Brian jumped down from the wing to join the conversation. "So, what was it, Colin?"

"Don't know, Mister Drummond. All the guns look good."

"Well they weren't good," he shouted, feeling the anger of a gun-fighter standing in the street with a gun jam. "All eight guns jammed in the middle of a friggin' fight. Something happened."

"Did you change anything on the guns between yesterday and today?" asked Gordon.

Jenkins thought for a moment.

Brian noticed several small groups near the other fighters. "Something important and common to all the fighters must have happened. Several of the others had all their guns jam as well," Brian added more calmly.

"The only thing that changed was the grease," Jenkins said.

"What grease?"

"The gun grease."

"Why?"

"The station armory blokes told us all, the Air Ministry directed us to use this new, improved grease."

Brian considered the possibilities. "Maybe something happens to it at altitude. We went through quite a few clouds and some rain, and it does get cold at altitude although we weren't really that high today. All I do know it my guns jammed. All of our guns jammed."

"I can't imagine anything like that," answered Jenkins. "We shall check the guns properly as well as see what Station and Group have to say about this, sir."

"I sure hope so. I don't want to do anything like that again."

"'Hunter,'" shouted 'Jackstay' Beamish from the dispersal tent. "Your turn with the spies."

Brian waved his hand in recognition. "Listen, guys. We're lucky we didn't get someone hurt up there. These fights are dangerous enough without adding to the excitement."

"Yes, sir," said Gordon. "We shall get her tip-top before you go again."

"I hope so, because I'm not particularly excited about doing that again," he repeated, as if the crew had not heard his pronouncement the first time.

"We'll get her right, Mister Drummond. We will," added Jenkins.

Brian saluted the men even though they had not saluted him. The mood of the pilots was not particularly good, as Brian passed through the group to complete his intelligence debriefing. By the time he returned, the others had tallied up the results. Of the twelve fighters that took off, three returned as Brian had -- jams on the first round. Four others had jams of one kind or another. All of them heard about the grease that appeared to be the problem. None of them were particularly enthusiastic about repeating the unfortunate episode. To everyone's relief, Squadron Leader Darling notified both Sector Control, as well as the Group Headquarters, of their experience and requested the squadron be removed from availability until the problem was remedied.

The three aircraft that experienced gun jams on the first round were towed to the gun pits for testing. The pilots were removed from Readiness status, but were not released for nearly an hour. They stewed over the experience as well as the frustration for being held at Dispersal. They were released at mid-afternoon. They returned to the Officer's Mess, and Brian thought about taking a nap.

"Mister Drummond," called the mess steward. "You received a telephone message about two hours ago." He handed Brian a folded piece of paper. He wanted it to be Charlotte Palmer asking him to come to Standing Oak Farm. The note simply asked him to call, Bushey Heath 2471 – Mary Spencer. He considered waiting until after his nap to call, but his curiosity took control.

The operator connected him promptly to the number indicated. "Bushey Heath, Two Four Seven One," said Mary softly.

"Mary, this is Brian. I got a note you called."

"Yes, Brian. I just wanted to hear your voice."

"What's wrong?"

"Nothing."

"It does seem a bit odd."

"Well, if you must know, I talked to John this afternoon, and he mentioned your squadron had some problems today."

The words struck Brian as rather strange.  They had only returned from the mission less than two hours ago and already people knew about the difficulty.  "It was nothing much.  We'll have the problem fixed in no time."

"I was just worried."

"Thank you, Mary, but no need to worry."

"As long as this damn war goes on, I shall always worry about you.  John is relatively safe in his bunker along with Churchill in his, but you are in the fight, so I worry."

Brian let silence occupy the space between them.  He wondered if that was the only reason she called.  Maybe she was going to try to talk him into meeting her again.  As much as he enjoyed his time with her, he told himself he had to say, no.  The sensation of her caress as well as the softness and warmth of her body did give him pause for thought.  He needed Anne, but he could never enjoy the pleasure of her comfort again.  Rosemary was probably at the other end of the country, and Charlotte was too important with whom to risk another mistake.  Maybe the risk with Mary would be worth it.  No, he chastised himself.  It was not right no matter how much he wanted the release of a woman's touch.

"Are you still there?" she asked.

"Sure."

"When are you going to be in London again?"

"I don't know."

"When will we have time together?"

"Mary, I do enjoy my time with you, but isn't this too risky.  Someday we are going to make a mistake we will all regret."

"The risk adds the spice, my dear fellow."

"Not that kind of risk."

"You are released early.  You could probably be in London within two or three hours.  Think of the pleasure we might enjoy."

"I know."

"Well?"

"Mary, I can't."

She laughed.  "I figured as much.  One never knows until one asks for what one wants.  Actually, I was truly calling just to hear your voice and make sure you were all right."

"Thanks, Mary.  I am perfectly OK.  No problems here."

"Excellent."

"I shall not occupy your precious time any more, dear love.  Please take care of yourself and be as careful as you can.  I don't want to lose you."

"You're not going to lose me," Brian answered, although he thought, she never had him.

"I do love you, sweet dear."

"And, I love you."

"Then, until the 'morrow, *adieu*."

"Bye," he said and hung up the handset.

Brian went to his room.  He sat on the bed to think about the conversation with Mary Spencer.  He liked, and in many ways needed, to let go of the tension and pressure of aerial combat.  He fell asleep with his uniform still on and his feet angled over the edge of the bed.  It was the image of Charlotte Palmer that stayed with him until unconsciousness took him.

—

*Monday, 26.August.1940*
*Oval Office*
*The White House*
*Washington, District of Columbia*
*United States of America*
*14:30 hours*

"On behalf of the American people, welcome to the United States, Sir Henry," President Roosevelt offered to the leader of the British Technical and Scientific Exchange Mission.

Sir Henry Tizard shook the President's proffered right hand, and responded, "It is an honor to meet you, Mister President.  Thank you for seeing me personally."

"I would not have it any other way.  Your mission is far too important to Prime Minister Churchill, myself and both our peoples."

"Thank you, sir."

The small group situated themselves.  Also in attendance for this meeting were Secretary of State Hull, British Ambassador Lord Lothian, Secretary of the Navy Knox, Secretary of War Henry Lewis Stimson of New York, who was also a prominent and influential Republican Party politician, along with his colleague 'Frank' Knox, British Security Coordinator Stephenson, and National Defense Research Committee (NDRC) Chairman Vannevar Bush. The ever-present Harry Lloyd Hopkins – assistant to the President – sat behind Roosevelt in a single chair next to the large desk.  Tea or coffee was served each attendee who wanted a cup.  The social cordialities were dispensed with in short order.

"If you would allow me to get directly to the point of this gathering," began Roosevelt, "what is the plan for this historic visit?"

Tizard scanned the group to ensure he was not going to speak out of turn.  "As I imagine you are aware, Mister President, I arrived a few days ago along with one of our team members Group Captain Fredrick Pearce.  We are serving as the advance team for the entire group.  I have also been led to believe you are familiar with the final list of topics and exchange items."

"Yes, I am, if the list has not changed in the last few weeks."

"The list has not changed since it was approved by the Prime Minister, the War Cabinet and the Defense Committee at the beginning of August.  The remainder of the British half of the exchange team along with all of the papers and cargo on that approved list are schedule to board the *Duchess of Richmond* in four days."

"Along with the Royal Navy crews for eight destroyers, if I am properly informed," injected Roosevelt.

"Yes, sir, that is my understanding as well.  The ship should arrive in Halifax, Nova Scotia, on the 6th of September, to effect delivery of those destroyers.  Arrangements have been made for the cargo to be promptly off-loaded and transferred by rail to a secure facility at the Washington Navy Yard at Anacostia.  A joint security team will protect the shipment from the ship to the storage facility at which time we will formally transfer responsibility to you."

"We are ready, are we not?" the President asked, looking to each of the Americans.  He received affirmation from each of them.

"I must say at this point, Mister President, your people have been most helpful and accommodating, and I expect will continue to do so during this process."

"I expect nothing less."

"Thank you, sir.  Lord Lothian and his embassy staff have helped us find lodging for the team as well as office space, both of which will be in the Shoreham Hotel for the time being.  As soon as we get the British contingent to Washington and settled, which we expect to be completed by Sunday the 8th, Mister Bush and Secretary Knox have graciously arranged for a joint team conference at a secure facility on Anacostia Naval Station, where we will begin the process of passing to our American counterparts the various items, associated engineering and scientific research on the list.  Once the general coordinating meeting is concluded, we, Mister Bush and I, expect the defined sub-teams to break-up and proceeded to their respective facilities."

"How many groups are there?"

"Eight, sir. The list has been divided into eight main groups – weapons, explosives, electronics and such."

"And everyone is matched up well?"

"Yes, sir, quite so. Not to speak for Mister Bush, but I believe we are both satisfied with the joint assignments."

"Sir Henry is correct, Mister President," Bush added.

"Excellent. How long do you expect this process to take?"

"By our joint plan, the initial exchange should be complete within a couple of months, at which time my charge will be fulfilled. I have briefed our team members to be prepared to remain at least until the end of this year. I think both Mister Bush and I see usefulness in selected experts remaining for the duration of the war."

"Excellent. To be clear, I acknowledge that you have asked for nothing in return. However, I have instructed all of our associated departments and agency that this will be an exchange program in the truest sense of the word. We will share openly our research and development work, and I expect you will see where we can contribute to our mutual defense technology. Both Prime Minister Churchill and I are in complete agreement regarding the significance and importance of this effort. I have made it clear I will not tolerate any obstacle or resistance to that end. I ask that each of you," he said, looking directly at Tizard, Lothian and Stephenson, "will convey to His Majesty's Government our most sincere gratitude for the generosity, trust and friendship this program represents."

"I can assure you, Mister President," responded Lord Lothian, "your kind wishes will be faithfully passed to the Prime Minister and His Majesty's Government."

"Thank you. Now, in related business, Mister Ambassador, I can inform you that the last of the nasty, nuisance, legal hurdles were dispatched with last week for the destroyer transfer we have been so diligently working on for the last few months. Secretary Knox informs me," Roosevelt said and received a head nod from the Secretary of the Navy, "the first flotilla of eight destroyers are scheduled to arrive in Halifax shortly after the *Richmond* docks, after completion of this sea trials. I must say, the refurbishment of those surplus old destroyers was not as far along as I had hoped, but I am assured they are fully armed with fuel, provisions and ammunition for all the weapons systems aboard."

"Thank you, sir. I will immediately inform the Foreign Office, Number Ten, and the Admiralty."

"Is there anything else we need to discuss?" the President asked.

"Yes, sir, if you will permit me," said Tizard.  He waited for the President to nod his consent.  "Prime Minister Churchill asked me to acknowledge the sensitivity of the United States Government, regarding the Norden bombsight.  We recognize and accept that the device is not a part of this exchange program or any other collaborative effort between our two nations.  As you may know, for several years now, our bomber designs have required space, mounting, power and control provisions for the bombsight be incorporated in every one of our designs to allow prompt incorporation when the day arrives that you are able to share the device with us.  We ask for nothing more than continued coordination with your defense establishment to ensure that compatibility is maintained.  As I am sure you can imagine, we do have a most pressing need."

"Henry, I think this is your area."

Stimson sat up straight.  "Yes, sir, Mister President.  We are prepared and committed to give the British government those assurances."

"Very well, anything else?"  The President gestured to each man and received a negative reply.  "Harry," said Roosevelt, glancing over his right shoulder, "please record the content of this historic meeting.  Again, Sir Henry, welcome to the United States of America.  We are honored to have you and your group as our guests and colleagues.  May your mission here be wholly successful."

"I am certain it will.  Thank you very much for your generous time and support." Sir Henry stood and took Roosevelt's extended right hand.

The men except Hopkins departed the Oval Office without further words.

Harry waited for the door to close.  "I wish I understood why that Norden bombsight was so important to the British."

"Harry, my fear fellow, it is a most impressive device.  A few years ago, I witnessed a most awe-inspiring demonstration of that device at Fort Benning, Georgia, during one of my extended stays at Warm Springs.  An Army Air Corps bomber repeatedly dropped a bomb into a twenty-five foot circle target from various altitudes and attack directions. So I understand and believe, it is the most accurate bombsight in the world.  The Army and Navy are both quite reluctant to give anyone access to the device, or the engineering and mathematics behind the sight, although J. Edgar Hoover believes the Germans already have the drawings."

"Is that so?"

"According to Hoover, but he has not been able to prove it, as yet."

"If the Germans have it, shouldn't we give the unit to the British?"

"If it was up to me, yes, I would, based on Hoover's suspicions alone, and I suspect he is correct. But, Stimson and his generals, and Knox and his admirals, are adamant in their denial of Hoover's suspicions and the necessity to close-hold that bombsight."

"You could overrule them."

"Yes, I could, but I will not. I need Knox and Stimson for the coming war, and I will not risk losing their support. After all, I am hopeful the unilateral generosity of this exchange program will soften the stance of the generals, so we can better help our British cousins. We need them to have every tool we can provide."

"Well, then, so be it."

The President and his assistant moved on to other topics and the rest of the day's agenda.

———

*Tuesday, 27.August.1940*
*RAF Middle Wallop*
*Middle Wallop, Hampshire, England*

The wake-up call came very early even though it was raining outside and the thick cloud cover made the morning seem much earlier than the clock indicated. Brian found himself in the same position he was the previous afternoon. The rain bode well for the day, but each event in the daily routine remained as it was on every fine day in the battle – wake-up, shower, dress, quick breakfast and transport to dispersal. This day was no different, other than the rain.

As they waited, they learned about the gun problems. The new grease apparently had a tendency to thicken at the cold temperatures of high altitude flight and especially with moisture in the vicinity. All the guns had been torn down, cleaned and restored last night. Crews worked until nearly dawn to complete the chore. The old grease was used this time. The fighters were ready, but they waited. There was activity to the East, and they did come to Readiness once in preparation for a launch to assist No.11 Group. The launch command did not come. They returned to Available status. While they waited, they learned that No.616 Squadron based at RAF Kenley lost six pilots, half the squadron, in less than two hours of aerial combat, drawing a very fine point on the dangers the RAF Fighter Command pilots faced during this deadly summer of war.

With the forecast predicting rain and general, broad cloudiness, and the RDF scopes nearly devoid of inbound German raids, No.609 Squadron and many other Fighter Command squadrons were released for the remainder of the day.  The prospect of an afternoon's and evening's relaxation from the air battle improved the mood automatically.  They were repeatedly reminded and warned by Squadron Leader Darling that they did not have a rest pass and had to be fully ready to fly and fight at dawn tomorrow.  Despite the warning, the small group decided this was the opportunity for an excursion to London. Brian toyed with the idea of making the journey to Standing Oak Farm, but quickly agreed to go with the group.  It was the right time for a meeting at Shepherds Pub in Mayfair.

The squadron pilots bounced into the Officer's Mess.  The truck would wait and transport them to Andover to catch the Southern Railways train into London.  "Mister Drummond," called the mess steward, as they entered the lobby.  "You received another telephone message.  A woman again, sir."  This time the others heard the announcement resulting in a variety of gibes about his love life.  How could Mary Spencer know he was heading to London?  Did she have someone out here feeding her information?

Brian looked at Jonathan and whispered, "Did you tell her we were going to London?"

"Who?"

"Mary Spencer."

"Bloody Christ, Brian.  Surely, you are not still seeing her."

Brian held his right index finger to his lips.  "No, but she called me yesterday.  She still wants to see me."

Jonathan looked around to satisfy himself no one could be listening in on their conversation.  "Is the call from her?"

Brian checked the note.  "No.  It's from Charlotte."

"The woman who saved your arse?"

"The one."

"So, does this mean you are not going to London with us?"

"No.  I'll go, but I'd better give her a call."

Jonathan left Brian and ascended the large staircase to get ready for the trip to London.  Brian looked at the note and went to the telephone booth. He waited for two other officers to complete their calls.  He dialed the number on the note.

"Standing Oak Farm," the male voice answered.

The voice startled Brian.  He had not expected to hear a man's voice in place of Charlotte's delightfully melodious tones.

"Hello?"

"Yes, yes.  This is Pilot Officer Drummond returning Mrs. Palmer's call."

"Yes, sir. She is out back. I shall fetch her. It should only be a minute."

"Thank you," Brian responded.

More than two minutes passed before Charlotte's voice came to him. "So good of you to return my call, Brian."

"It's nice to hear your voice, Charlotte. Is everything OK?"

"Quite all right, I should think. I called because of this award ceremony day after tomorrow. I was frankly a bit stunned when I received the invitation, and even more so when I learned the importance of this award. Brian, I am not comfortable with this sort of thing."

"To be quite honest, Charlotte, I'm a little apprehensive as well. The two pilots who most influenced my flying both received the Distinguished Flying Cross in the Great War, and I don't consider myself to be in their league. You deserve the George Cross, or at least I think so. I owe you my life."

"Nonsense, Brian. I just reacted."

"Whatever you did, you saved my life."

"But, the George Cross."

"I think it is great."

"And from the King."

"Yes, quite an honor, I should think."

"I am rather scared, Brian, which is why I called you. Can you accept the award for me?"

Brian thought about her request. "Do you think that is proper?" he asked, sounding very British.

"I am scared."

"Of what? He's just a man."

"He is the King."

"He's still just a man."

"Maybe to you Americans, but he is royalty. He is the Crown."

"Charlotte. I'll do what you want me to do," Brian paused, as the possibility of not seeing her crept into his thoughts, "but, I really think you should. This is important for you."

"Will you be there with me?"

The question flushed Brian with the warmth of her need. He liked the feeling. Maybe it was a shallow feeling, but it made him feel good, nonetheless. "Yes, I'll be there."

"Good, then, I should be fine. You will be the only person I know, I am quite sure."

"Your family won't be there to see you receive your award?"

The silence on the other end told Brian he had touched a sensitive spot.  He did not know much about her background other than her husband had died at sea.  Maybe she had problems with her family?  Maybe she was embarrassed to have them meet him?

"I have no family left," she said with solemnity.

"I'm sorry, Charlotte.  I didn't mean to say anything wrong."

"You had no way to know, Brian.  I was an only child.  My mother was killed in a ferry accident.  My father and uncle were killed in the trenches during the Great War."

"I am sorry."

"Thank you."  Silence filled the space between them, leaving Brian without anything comfortable to say.

A knock on the window of the telephone booth door startled Brian.  Jonathan pointed at his watch, and then gave him the chocks-out signal telling him the group was getting ready to leave.

"What was that?"

"My friend knocked on the door.  He's telling me I've got to go."

"By all means," she said with a tone that told him she was not ready to end their conversation.

"Are you OK?"

"Quite."

"Then, you'll be here on Thursday?"

"Yes, Brian.  I'll be there on Thursday.  Just make sure you are there.  I want to see at least one friendly face."

"I will."

The telephone conversation ended without the tone he would have liked, but at least he would see her in two days.  Brian ran upstairs to his room, changed his shirt and tie, grabbed his tunic and brimmed hat, and ran back downstairs to the waiting truck with a load of impatient pilots.  As they rode up to Andover, Brian gave Jonathan, overheard by the others, a quick sketch of the telephone conversation.

During the train ride into London, the group had to split up into several compartments.  Jonathan and Brian took one, sharing it with three older women and a much older man.  Jonathan slept a good portion of the journey leaving Brian to his thoughts.  He wished he had been able to call Jeremy Morrison to meet him at Shepherds, but the call from Charlotte took all the available time.  Maybe he could call once they arrived in Mayfair.

They arrived at Waterloo Station, took the Underground to Green Park Station and walked into the Mayfair District.  The early afternoon hour did not stop respectable numbers of fighter pilots from gathering for refreshment and camaraderie.

"I shall have a few beers with the lads, and then I have a date with Linda," said Jonathan.

The pronouncement brought the persistent memory. Jonathan met Linda through Anne, but Linda had not been caught up in the spy ring affair. Jonathan still had Linda, while Brian had lost Anne.

"Sure," Brian answered.

"Right, then, let us have another pint of bitter."

The laughter, joking and frivolity of Shepherds Pub made everyone's mood better, even Brian's. As was always the case, the pilots shared tales of stupidity, valor, humility and courage in the air. There was strength in the experiences of others. Indirectly, they all learned from the shared experiences.

True to his word, Jonathan departed prior to the evening. Brian wished his best friend an enjoyable evening. He did not expect to see him until the next morning. The camaraderie did not skip a beat. Brian began to feel the soothing effects of his third pint of beer. The death and destruction of the war disappeared from the famous pub near the center of London.

"Sailor," someone shouted.

Brian turned toward the entrance of the rustic, old public house. Searching through the mass of RAF blue, Brian eventually caught a glimpse of Squadron Leader Adolph Gysbert 'Sailor' Malan, DSO, DFC, Commanding Officer of No.74 Squadron. He could not see Jeremy or any other familiar face. He returned to the conversation flowing around him. As he finished his third pint and made his way toward the bar, a hard slap on his back caused him to stumble into someone else. He turned to see the culprit.

"So, where have you been, 'Hunter?'" asked Flight Lieutenant 'Mud' Morrison.

"Fighting a war, same as you."

"So I hear."

"Yeah, what?"

"I hear you're now an ace, and you have a DFC for it."

"Not yet," Brian shouted over the din.

"Only for want of a few days."

"Yeah."

"I have been meaning to call you."

"Likewise."

"What is it they say in your country . . . let's blow this pop stand?"

Brian laughed. He had heard the phrase in the movies or on the radio, but he had never heard the words used in real circumstances in Wichita. He nodded his head.

Jeremy led Brian through the mass of fighter pilots and passed some of the most renown and revered names in Fighter Command. A light drizzle greeted them in the much cooler, fresh air. Jeremy waved for Brian to follow him, which he did. A small cafe with few remaining patrons attracted Jeremy's attention. They found an empty clean table and did not wait to be seated. The waiter responded to Jeremy's signal for two pints of beer.

"When was the last time we talked . . . a month ago?" Morrison began, but did not wait for an answer. "How have you been?"

"About the same as you, I suppose."

"I'll tell you with all candor, I loved Virgin," he paused with his eyes on some spot on the table. Brian always thought Jeremy's abbreviation of Virginia North's name was rather odd. ". . . probably as much as you loved Anne. It was such a tragic waste of exceptional womanhood. I miss them."

"Yeah."

"I would guess you have shaken off the loss."

"Not entirely."

"Me either."

They both sat quietly drawing several times on their beers. It began to rain harder, adding peacefulness to the evening.

"Have you learned any more now that they are gone?" Brian asked, the one question he held since the fateful day.

Jeremy took another drink of beer. "To be blunt, they both serviced numerous generals, admirals and diplomats. They apparently managed to obtain substantial amounts of very sensitive information about our armed forces, our defense system, and from what I have been told, critical Foreign Office information. They passed the information to the Germans."

"Why?" he repeated the same question for the umteenth time.

"I don't know. As I told you before, I tried to get into the prison to talk to Virgin, but they were all forbidden any visitors. From the severity of their sentence and the harshness of their treatment, my guess is they passed some very damaging information."

"Yeah, but why? Why did they betray their country like that?"

"It is the same question, and there are only the same answers. I even used my brother's peerage to find out more, but to no avail. Scotland Yard and the counterintelligence bastards locked this affair up tight," he paused to take another drink. "I wish I knew the real reason why. It would certainly make acceptance easier. She was such a good woman, so was Anne."

"Yes, she was."

"You know, Brian, I am sorry I got you into that sordid mess."

"I'm not," he said quickly. "As much as it hurt my ego when you got us together, she taught me so much about life. I'm glad I knew her. My greatest regret is losing her." This time Brian paused to take a drink. "I never did like what she did for a living, but I accepted it just to be close to her. I've never felt so relaxed around another human being. I miss her. I miss her a lot."

"You are not alone there."

"How's flying?"

"Nice way to change the subject. *Garçon, deux pintes supplémentaires, si vous plaît.* About as good as you. Well, not really. I am not an ace, yet, so it certainly isn't as good as you. You have become very much the fighter pilot I knew you would. You have an exceptional touch, Brian – a gift. It is going to rest on pilots like you to keep our shores free of Germans."

"And you."

"If I was only half as good as you."

"I don't need that. I learned a lot from you."

"Balderdash."

The conversation turned to much lighter subjects as they ordered and ate a nice meal. They returned to Shepherds and the mounting raucousness of the pub. The No.609 Squadron pilots along with others from the West and North departed to catch the last train out of London before tranquility began to return to the famous pub. The train ride back to Andover passed quickly through sleep. They arrived at the RAF Middle Wallop Officer's Mess in the wee-hours of the morning to find the few hours of sleep they could before the morning wake-up call.

———

# Chapter 9

A man can be destroyed but not defeated.

-- Ernest Hemingway

*Wednesday, 28.August.1940*
*RAF Manston*
*Ramsgate, Kent, England*
*16:15 hours*

## Week 8

The Prime Minister's visit to this place symbolized the significance of the current struggle as well as the true test of British resolve to hold the Germans at bay. The location of the Fighter Command base at the mouth of the Thames Estuary and the eastern most point of Kent brought unwanted attention to the minor facility. It had been attacked more than any other location, and yet, it had no permanently assigned squadrons. When the local commander decided to abandon the beleaguered base two days ago, Churchill reacted strongly, decisively and with considerable emotion.

The commander was relieved of his assignment. A new commander with very clear instructions was installed. The newspapers had notified the citizenry of the incredible pounding the hapless air base had taken during August and the commander's decision to abandon the site. The potential public reaction to the news shaped Churchill's response more than anything, and he knew it. The Prime Minister knew more than most that the key to victory in the face of daunting odds, overwhelming military might and dark prospects lay with the attitude of each Briton. If they believed they could prevent the Nazi invasion, they would muster up the wherewithal to succeed.

The automobile journey to RAF Manston had been a silent, brooding one for Winston. Low clouds kept the temperatures cool most of the distance and receded to the water's edge by the time of his mid-day arrival. The Prime Minister ordered the car stopped before they reached the station headquarters building. He stepped out of the car. The Prime Minister of Great Britain, dressed in the uniform of an RAF air commodore, stood solemnly with his head bowed. General Ismay, as usual, accompanied the British leader.

The devastation proved to be more than his imagination could conjure up based on the reports he read in Whitehall. Not one building in sight remained undamaged. More than a few had been reduced to a pile of displaced rubble, unrecognizable as the remainder of buildings. The heavy smell of smoke and burnt explosives saturated the air making every breath a struggle. The grounds of what must have been the aircraft landing area reminded Churchill of the chewed up, pocked, blood soaked fields of Flanders, but this was England.  Anger began to fill his veins as Winston walked among the still smoldering ruins out into the cratered field.  The sandbagged bomb shelters with their trenches made the scene even more graphic and reminiscent.

The new station commander, a wing commander whose name he did not remember, nearly ran toward them and saluted.  Both Churchill and Ismay returned the salute.  The officer appeared quite agitated and out of breath.

"Sir . . . sector control . . . reports . . . several inbound raids.  I should . . . encourage you . . . to join me . . . in the command shelter," the wing commander said with his chest heaving for air between words.

"Nonsense, young man.  We shall be perfectly safe," Winston replied, as he walked farther out onto the former landing area.

Ismay, the wing commander and several other officers followed their leader.  Churchill stopped numerous times to survey the bomb damage as well as point out evidence of unexploded ordnance to the not-so-curious entourage.  The distant drone of German bombers preceded the air raid siren by several minutes.

"Prime Minister, please," said the wing commander with less labored breathing, "we have been bombed everyday for the last fortnight, and surely this day will most probably be no different.  I implore you, please seek shelter."

Churchill turned in an instant to face the confused officer.  "Sir, this is our land, our beloved England.  No damnable German will move me from it alive."  Winston stood staring at the thin, youngish looking, shocked RAF officer.  "If you desire to seek shelter, then do so immediately.  The same for the rest of you as well," he said, waving his arms to the other men.  All but Churchill and Ismay ran for the bomb shelters.

As the grotesque groaning of the bombers grew louder, Churchill decided to light one of his prized, *Romeo y Julieta*, Cuban cigars and placed his fists on his hips with his eyes skyward.  The bombers approached at relatively low altitude making them plainly visible.  Even the Prime Minister could recognize the distinctive shape of the whale-like, He111, twin-engine, medium bombers – about twenty of them.  Above and behind them, the buzzing of the escort fighters mixed into the scene.  Within minutes, Churchill and Ismay watched, and listened to a squadron of Spitfires jump into the German fighters as a squadron of Hurricanes raced head-on into the bombers.  The altitude provided a most exquisite view of the engagement, Churchill told himself.

The characteristic whamph-whamph-whamph of exploding bombs just to the south and walking closer to them convinced General Ismay to place a hand on his leader's shoulder.  "Prime Minister, don't you think we should join the others?"

"Nonsense. Can't you see? Our young flyers are winning this battle. The bombardiers have lost their nerve and dropped their heinous loads early. The bombs are falling short, and I would not miss this exquisite exhibition for the life of me."

The battle continued not quite directly above them, as the bombers struggled to turn and retreat. Two bombers went down. One Hurricane was smoking badly. They watched the pilot parachute from his stricken machine. A Spitfire spiraled down with the tip of its right wing missing; the pilot did not make it out of the damaged aircraft. A German fighter exploded in mid-air. Another one, streaming black smoke behind it, weaved trying to avoid the two pursuing Spitfires. The valiant German lost his struggle, as his airplane rolled slowly, and then plunged into the water. Two more German fighters trailed faint white smoke, but managed to escape. The sadness of bearing witness to the battle overhead was displaced by the success – the bombers missed their target.

As if to pay tribute to the beleaguered auxiliary fighter air base, a section of three Spitfires descended from their victory, ran toward the airfield at grass-top level and very high speed. The sunlight flickered off the large propeller disks on the nose of each aircraft. The three exquisite machines exuded power. Churchill jumped with joy watching the sleek elegance of the British fighters heading slightly to his right toward the center of the RAF base. The glorious sounds of the triumphant fighters brought streams of tears to his eyes as they pulled up with a pronounced growl of the engines, split into a starburst, rolled twice each and rejoined on their leader for the return to their base.

The Prime Minister of Great Britain, the Minister of Defense, First Lord of the Treasury and chief cheer leader for free Europe, the Right Honorable Winston Spencer Churchill stood crying, laughing and shaking his fists in the air. "We're going to beat these bastards. I shout for all the world to hear me. We are going to place that vile menace in an unmarked grave where he belongs."

As the scene settled, Churchill wiped the tears from his face, gathered his composure and turned to see the beaming face of General Ismay who in turn held both his fists out toward his leader as if to make a secret salute to a brother in arms. Churchill saluted Ismay who returned the salute. The two men shook hands.

"You truly believe we are going to win this fight, don't you?" asked Ismay.

Churchill smiled. "With terribly young men like those," he said, waving his hand in the direction of the vanished British fighters, "who can celebrate with enthusiasm after mortal combat, how can we not win. Yes, I do sense we have turned the corner, although we have many dark days ahead.  We owe these young men the honor of their victory, and they shall have it."

"Yes, sir."

The two men walked back toward what was left of the facility.  Air Force personnel emerged from the bunkers looking around to see any further dislodgment that might have occurred.  Winston noticed the strange almost reverent expressions on the faces of the young men, boys actually, who watched him walk among the scattered bricks, broken boards and tangled wires.  He tried not to respond to the faces of those around him, to do so would be to acknowledge their reverence.  As he moved around the various piles of debris, Winston fought to suppress but could not avoid the memories of all those faces beside him in the trenches of Flanders fields 24 years earlier.

As an unseated First Lord of the Admiralty after the disaster of Gallipoli, Winston defied his supporters and family, and joined the prime of British manhood in the trenches of the Great War.  The fall from the highest post in the Royal Navy and one of the traditionally most senior positions within His Majesty's Government to a lieutenant colonel of infantry in the British Army had been precipitous and many thought fatal.  The experiences on the battlefields of the War to End All Wars etched themselves deeply into his psyche and character.  He could never forget.  The young faces around him at the battered RAF base and the memories of the trenches made the 66-year-old man feel forty years younger.  They gave him a vitality he had not felt in many years.  The isolation, ridicule and ostracism of those lonely wilderness years, as the sole voice in opposition from the backbenches, evaporated among the men who now stood in defense of Mother England.

During the return trip to Whitehall, Winston transformed the energy he absorbed at that lonely spot in Eastern Kent into the will of a nation.  General Ismay resorted to a small notebook and pencil to capture the directives of the Minister of Defense.  All craters must be filled in and the landing areas returned to acceptable condition within 24 hours.  No base, aerodrome or facility would be abandoned due to enemy bombardment.  The Royal Navy must complete work as soon as possible on HMS *Erebus*, the unpowered, heavily armored, long range, naval bombardment vessel, and deploy her promptly for counter-battery fire against the German guns of Cape Gris Nez.

The conversion of Churchill's newly infused energy enabled him to regain a stoic, determined exterior.  Only General Ismay had witnessed his boyish enthusiasm and excitement, and he had been sworn to secrecy.  The inspection of the now operational Cabinet War Rooms bunker under the west end of the New Public Offices building simply did not fit with the afternoon's episode. To Winston's regret, he had to return to the mundane business of leading His Majesty's Government, the British people, the Commonwealth and the free world in the mortal struggle against oppression.

———

*Wednesday, 28.August.1940*
*Cosmos Club*
*2121 Massachusetts Avenue, NW*
*Washington, District of Columbia,*
*United States of America*
*20:00 hours*

"Welcome back to the Cosmos, Mister Bush," greeted the elderly male host. "We have your requested private room ready for you."

"Thank you very much, Mister Phillips."

"Mister Morgan, please escort Mister Bush and his guest to the Adams Room."

"Yes, sir."

"Thank you again, Mister Phillips."

"By all means," responded Phillips.

Sir Henry Tizard and Vannevar Bush followed the assistant host up one circular flight of stairs and down a long corridor to the last door on the left. The well-appointed room featured an eight-place dining table set for two opposite from each other at the far end from the door. Two, large, John Frederick Kensett, landscape paintings dominated the long walls of the room. A large, floral arrangement of red roses and white carnations occupied the center of the table. The large window had no blackout curtains, and the installed, elegant curtains were undisturbed in their bloused, open position. Sir Henry noted with some confusing irritation the plethora of bright, cityscape lights beyond the window – a dramatic contrast with the reality he left in London.

As they were seated, Bush noted Sir Henry's attention on the window. "We are not yet at war."

"Quite so. Conditions in London are substantially different. We look forward to the day we can turn our lights back on."

"I am sorry that your countrymen must endure such abuse. Hopefully, our work together will bring that day of celebration closer for you, and for all of us."

"Indeed."

Although they had not yet discussed any sensitive information, they both paused as the sommelier served a nice, French, vintage, cabernet sauvignon and their waiter served a savory French onion soup.

"After our meeting with the President Monday afternoon, I thought a more social venue might assist our relationship."

"A nice touch, I must say."

Bush did not skip a beat. "What is the latest news from England?"

"The message I received this morning through the embassy by coded radio informed me the listed items are being loaded aboard ship, the team is assembling in Liverpool, and the *Richmond* will depart Liverpool on schedule this Friday evening, and should arrive in Halifax on Friday, the 6th of September."

"Excellent. We have coordinated with the Canadians. Once your team and cargo are loaded aboard the railcars, the American security team will protect it and ensure all of the cargo is moved safely to the secure facility at Anacostia. I have also received confirmation from each of the sub-team sites that they are prepared to receive both the cargo and personnel whenever we set the transfer in motion. I think we are ready for their arrival."

"Yes, I agree, although I must confide in you, I cannot claim the same on our side. I was rather disappointed our embassy had not done a better job arranging suitable office space for my team. I took it upon myself to contract a suite at the Shoreham to include the replacement of the hotel furniture with desk, chairs and telephones. We will use that temporary office until a proper business office can be established."

"I can certainly help there, if you need me to do so, Sir Henry."

"Thank you, Van. I expect to have this nuisance settled in short order."

They waited again for the next course. They both chose a generous cut, Kansas City Strip steak, with garlic green beans and butter-roasted new potatoes. Sir Henry had not tasted beef in many months, so he thoroughly enjoyed his first few bites.

"Food rationing has taken its toll on all of us," Tizard said, after a few minutes of pleasure.

"I can only imagine and I certainly hope it does not come to that in this country."

"I hope so as well."

They continued to eat in silence as Bush allowed his guest uninterrupted enjoyment. Vannevar Bush waited until Sir Henry Tizard had consumed half his meal before he spoke again.

"If your schedule permits, I would like to escort you to each of the eight sites where the sub-teams will be meeting to meet the men involved and to familiarize yourself with the sites, and the work they are performing.

I believe these visits will help us integrate the various teams when we all meeting for the general meetings in Washington, next month."

"Yes, yes, absolutely.  That would be most helpful, and I do agree completely.  I stand at your service, Van."

"Excellent.  Now, if I may," Van said and waited for a confirmatory head nod from Henry, "one topic we have not discussed with sufficient detail, as yet, is the matter of uranium explosives."

"Yes, certainly, we have included the issue in the general explosives line item.  We did not want to draw unwanted attention to the subject."

"Quite appropriate, I should think.  I understand some in your government are not happy with our intensity, shall we say, regarding nuclear research."

Sir Henry smiled as he searched Van's eyes.  "I can only presume you have heard directly or indirectly from Mark Oliphant."

"Indirectly."

"Yes, well, Mark has found religion, as they say."

Marcus Laurence Elwin 'Mark' Oliphant was an Australian born, Cambridge educated, disciple of Nobel-laureate Sir John Rutherford.  He had been an enthusiastic and energetic member of the British MAUD Committee (Military Application of Uranium Detonation) from its inception and had already made two trips to the United States to confer with American colleagues, regarding the potential for nuclear explosives.

"To be perfectly candid and direct, Van . . . Mark has voiced his apprehension, and I surmise the unspoken concerns of his British contemporaries and his American counterparts, regarding the leadership of your uranium committee."

"Lyman Briggs?"

"Yes, to be frank."

Lyman James Briggs was the Director of the National Bureau of Standards for the federal government.  President Roosevelt personally selected him to lead an exploratory committee of physicists as a consequence of the 1939 letter from Doctor Albert Einstein, expressing serious concern about the German government nuclear research and the ominous specter of the Germans producing a bomb of such enormous power.

"My turn to say, yes, well," Bush responded and chuckled.  "I am currently a little too distant to have direct knowledge.  You may not be aware, but Briggs and Roosevelt go back a few years.  The President himself picked Briggs for the assignment, so I think it safe to say, he was a political appointment, and thus not easily dislodged."

"Most unfortunate."

"Please share with me the concerns of your government," said Bush.

"I am not sure that is appropriate, given the information you have offered, Van."

"Sir Henry, we will be dealing with many sensitive and sticky issues in the coming weeks and months. Much of the success of this program will rest upon the trust and confidence you and I have in each other. I suppose this is our first meager test of that trust. I can only assure you, I will not compromise your confidence in me or engage directly until you wish me to do so. Further, if you so stipulate, what is shared between us will remain between us until such time as you allow me more latitude."

"Very well then, I shall so stipulate, this must remain between us for the moment, and I shall endeavor to explain why. Prime Minis . . ." he stopped, when the waiter entered to serve dessert – a rather tasty, peach custard crisp.

"Please leave a nice bottle of sherry, and that will be all." They waited for the sommelier to serve their sherry and two opened bottles. The door closed behind him.

"As I was about to say, Prime Minister Churchill gave me numerous instructions. One of those directives was to build relationships that will hopefully endure for the duration of these hostilities and beyond."

"Wise words, it seems to me."

"Yes, so you can appreciate my apprehension. I am a guest in your house. It is rather tactless for a guest to criticize a household member of his host, and I am quite mindful of Mister Churchill's words."

"I thank you for your sensitivity in such matters, Henry . . . may I call you Henry?"

"By all means, Van."

"Very well, then. From the President's charge to me, and it sounds like Mister Churchill's charge to you, perhaps we should think of ourselves as brothers, not just cousins, in a much larger family that must defeat the fascist as quickly as we can. We must be able to discuss issues promptly, directly and frankly, if we are to be successful in this endeavor."

"Well said and agreed. In the spirit of that brotherly relationship, I have only discussed this question with Mark a couple of times in the last few months, so I am by no means fully conversant. I am not speaking on behalf of His Majesty's Government, and only as a scientist who agrees with his friend. As Mark has expressed his concerns to me, the issue is Mister Briggs appears to be fully constrained by the plodding pace of bureaucracy with no sense of urgency. He seems to have little interest in the important topic before his chairmanship. Mark has conveyed, from his perspective, similar concerns by more than a few of his American counterparts. While our MAUD Committee has not yet completed their work, I know they are unanimously convinced the conclusions reached by Otto Frisch and Rudi Peierls in their memorandum of last spring are correct and perhaps even an underestimation of the enormous energy potential of nuclear fission. To be blunt, the resources of His Majesty's Government are being consumed at an alarming rate. We need the United States to pick up the lance. This task demands the capacity and industry of your country. Lastly, let it suffice to say, we, and by this I mean the British government, feel the German threat in very personal, intense and expansive terms. Mark is the most outspoken of our scientists, but he speaks for most of us, even though perhaps more directly than we would prefer."

"Thank you for sharing your perspective with me, Henry. I shall be equally candid. The President and his administration walk a very fine jagged line between the prevalent isolationist mood in this country and what I believe is his realistic appreciation of the world situation and our place in it. I have not discussed the Briggs issue with the President or anyone else. I know Lyman is a capable administrator and a competent scientist. While I do not have a polling of our scientific community or even our uranium committee regarding Lyman's leadership, I do know Mark is not alone among those scientists on this side of the Atlantic. The best I can do is look for opportunities to move us in the correct direction with sufficient urgency."

"I am afraid I could not ask for more at this juncture."

"I expect as our sub-teams break up to bore into their assigned subjects, perhaps that will open the opportunity for us to address our involvement and pace of research. If, or once it becomes a joint issue, I should be able to address the question with the President. He is a very smart man. I know he will help us do the right thing, when the time comes. I surmise you believe the time has already passed."

"Yes."

"If you are asking me to speak to the President about Lyman Briggs today, I will do so."

Sir Henry Tizard considered Van Bush's statement. "No. I must trust your judgment in such matters. I am satisfied that you have heard me out in friendship, and you are aware of and sensitive to our concerns. I suspect you will find the correct moment in short order. I just pray the might of the United States can be brought to bear on this task sooner rather than later, and . . . the Germans do not beat us to the punch."

"I shall do my best, Henry.  Now, I would suggest we call it a night."

"Thank you so very much for your hospitality, and for the delightful supper and conversation.  I look forward to many productive days ahead."

"You are most welcome.  I suspect this may be out last private discussion.  The Army and Navy are making it very clear they expect to be directly represented at every level, including ours."

Tizard laughed.  "It seems the military share many attributes regardless of the uniform or language."

"Yes, indeed."

The two men departed the Cosmos Club.  Bush had his driver drop Tizard off at the Shoreham Hotel before returning home to rest before another busy day.

—

*Thursday, 29.August.1940*
*RAF Middle Wallop*
*Middle Wallop, Hampshire, England*

The morning began for Brian like most of the previous mornings since the real war flared to life in early May, except this time the wake-up call came two hours later.  The shower helped him step from the fog of deep fatigue-induced slumber into the morning's brightly defined alertness of calm concentration.  At the breakfast table, the aromas of fresh bacon, eggs cooked any way you liked them, fried potatoes with extra pepper and this morning's sliced fresh tomatoes made the day feel good.  The smells and tastes often brought memories of Sunday morning breakfast in Wichita, Kansas.  Brian had learned months ago to not fight but control those images of home.  Homesickness carried a lack of concentration that could prove to be fatal in a flash.  The luxury could not be afforded at any price.

Pilot Officer Roland 'Boxer' Stockard joined the gathered pilots late but still within the designated serving time.  He wore a broad smile and waited until he sat down with his napkin in his lap before he spoke.  "Did anyone hear?" he asked but did not wait for an answer.  "Last night, 'Cat's Eyes' Cunningham bagged his third Gerry."

"Don't say," interjected 'Red' Burns.

"Flight of three bombers unescorted heading toward Manchester or Liverpool, they say.  'Cat's Eyes' sneaked up behind them, fired a load into the left wingman and exploded the bastard.  Flames nearly blinded the crew, but he managed to use the light to find another one.  He pumped the rest of his bullets into the sorry bastard, but could not bring him down."  'Boxer' started laughing as the tired and still half asleep pilots listened.  "Apparently, when he fired the first rounds, it must have startled the Gerry pilots. They started gyrating trying to evade and nearly ran into one another. Can you imagine? Flying along fat, dumb and happy thinking you are all alone in the dark sky, and all of sudden bullets start pinging all over your machine." He continued to laugh although no one joined him. "Friggin' bastards must have wet themselves." Stockard finished his story and drifted back into himself. The other pilots finished their morning meal in silence.

Squadron Leader Darling, who usually did not join them for breakfast since his wife had moved into their cottage across the carriageway from the dispersal tent, struck his water glass several times for attention. He waited for everyone's focus. "As you know, today we have been granted a half day's respite to honor our brethren. 'Jackstay' and 'Hunter' shall receive their much deserved Distinguished Flying Crosses for achievement in the air. More notably, 'Hunter's savior receives the George Cross for her courage in risking her own life to save his." Several pilots complemented Darling's words with, here-here's. A clear image of Charlotte Palmer's delicate face and gentle manner came to Brian. "Of course, we are certainly grateful for her selflessness, as is the nation. For this, we have the unusual privilege of hosting the King who will make the awards to Mrs. Palmer as well as our two mates. Air Vice-Marshal Brand will also be in attendance along with the press – the world press, so I have been told." Grumbles and grunts signaled their disdain for the unwanted attention. "I must encourage you to be on your best behavior." More groans of disappointment mocked the admonition. "No gratuitous comments, if you please." More groans. "Needless to say, it is not everyday blokes the likes of us get a visit from the King. Let's be up to the occasion."

With the announcement made, they were given several hours to relax before they would meet at mid-morning at the Operations building that had been freshly repaired and painted. The King was expected at 10:30. He would make a short speech, pin the awards, and then depart. Roger Beamish and Brian Drummond were expected to make themselves available to the press. Charlotte, as a civilian, would be allowed to make her own choice regarding the press. The proper etiquette in the presence of the King was discussed in several forms. The No.609 Squadron pilots lounged in the large sitting room. Several smoked cigarettes. Brian felt the need for fresh air.

"Where are you going?" asked Jonathan Kensington.

"I thought I'd walked down to the field."

"Why? It is more than an hour until we must be down there."

"Charlotte might come early."

"Oh, my."

"It's not like that, Jonathan.  She doesn't know anyone up here.  She has no family, and she said she was a little scared."

"Would you mind if I go with you?"

"No.  Of course, not."

The two pilots left the Officer's Mess without fanfare or notice.  The sky remained overcast.  The relatively cool, humid conditions kept everything damp.  They walked across the A343 carriageway, gave the gate guards a cheerful greeting and made their way through the orthogonal, narrow streets of the air base.  Several people, enlisted and officer, male and female, who knew or recognized Brian, said hello and offered congratulations for his impending award. Jonathan enjoyed the recognition for his best friend more than Brian did.  Ever since Squadron Leader Darling's announcement of the award last week, Brian felt uneasy.  Both Group Captain Spencer and Malcolm Bainbridge had been awarded the Distinguished Flying Cross, and Brian still did not consider himself to be a peer of theirs in skill, stature, accomplishment or any other way.

Brian and Jonathan reached the Operations building and control tower, and stood near the left front corner looking out over the large, green field with several dark brown circles that had been bomb craters.  The grass had been freshly cut, probably several times in the last few days.  Except for the pockmarks of the filled in craters, the large grass field could pass for a golf course.  Only the microphone stand, a dozen yards or so in front of the building, was out of place in the scene.  A single Spitfire from No.609 Squadron and a single Hurricane from No.238 Squadron had been positioned on either side of the microphone stand with their noses pointed inward toward the building.  The surrounding hills with the rich green of the trees and the colors of the few buildings in sight gave Brian the impression of idyllic splendor.  He liked this country.  He liked the fact that the country stayed green all year long except for the plowed fields and the deciduous trees that lost their leaves in the fall.  Brian had grown to love the country and the people, and more importantly he felt very protective of this beautiful place.  Standing with Jonathan looking out across the countryside, Brian finally connected with the words of Malcolm Bainbridge when he tried to explain to him the rewards for the risks.  Brian felt more than ever Malcolm's spirit with him, smiling down upon him and warming his heart.  He knew this was right.  He also knew his parents would understand once he was able to explain these feelings to them.

The other pilots began to arrive.  The time approached 10:00.  Several officers from the Air Ministry as well as Wing Commander Worly, who had the task of making sure everything was in order, began to fret about the arrangements.  Once all the squadron pilots had arrived, they were briefed several more times on protocol, procedures and the sequence of events. Other people began to arrive including numerous civilians in brown, gray and odd looking suits with large press banners hanging around their necks. The press and other visitors were asked to remain in a designated area until after the ceremony. Brian began to ask himself, where was Charlotte? Had anything happened to her?

Just as the urge grew toward his threshold, to ask someone to find out where she was, Charlotte Palmer appeared from an Army sedan. An Army major and an Air Force squadron leader escorted her toward the gathering of pilots. She wore a simple, but well-tailored, light blue, woman's suit with matching brimmed hat and a white ruffled blouse that gathered around her neck. Her hair was pulled up under her hat. The bright red lipstick amplified the fullness of her lips, and her eyes sparkled.

Brian's heartbeat jumped a few notches as he watched her walk carefully across the grass toward them. The muffled comments from some of his colleagues told him the effect her appearance had on the others. Brian started to break from his position to greet her, but stopped both from Darling's admonishment as well as the half dozen other officers who approached her to complete the final preparations.

Squadron Leader Darling ordered the officers to form their ranks, and then the enlisted men of the squadron formed in rank behind the pilots. Flight Lieutenant Beamish and Pilot Officer Drummond stood side-by-side in front of the entire formation facing toward the south away from the rest of the squadron. Charlotte was supposed to stand on the other side of 'Jackstay' Beamish, but Brian had not seen her since they were placed in formation. Maybe the best he could hope for was a brief glimpse of her as she arrived. Hopefully, she would not disappear as quickly as she appeared. He had not seen her in more than a week and had only been able to talk to her for that short time two days ago. Brian wanted the connection between them to be stronger than it was.

From the left corner of his field of vision, Brian saw the blue of her dress suit. The Army major walked on her right between them. As they passed in front of him, Charlotte slowed slightly to look behind the major and directly at Brian. She smiled and winked at him. Her face and eyes radiated warmth and energy on the cool summer overcast morning. Charlotte moved to her position on the far side of Beamish.

With Charlotte in position, the sequence of events stepped quickly. On cue, Squadron Leader Darling brought the squadron to attention. The rustle of people and papers moving behind them signaled the commencement. Everything became quiet, and then a few camera flash bulbs went off.

"Ladies and gentlemen, His Royal Highness, King George the Sixth," an announcer said over the loudspeakers.

The handsome, distinguished looking man in the uniform of a Marshal of the Royal Air Force walked to the microphone.  Although Brian had never seen photographs or paintings of him in an Air Force uniform, the refined features of his face and vibrant energy in his eyes left no doubt who the man was.

"Ladies and gentlemen, I shall keep my remarks brief this morning.  We all have duties that await our attention, and the present struggle demands every hand to the task, including these exemplary pilots and their ground crews.  We are here to honor three individuals for their contributions in defense of the realm.  Flight Lieutenant Roger Beamish from Scotland and Pilot Officer Brian Drummond from the United States of America, both are fighter pilots with Six Oh Nine Squadron here at RAF Middle Wallop.  They are to be awarded the Distinguished Flying Cross for masterful skill, courage and accomplishment in the air having achieved more than five aerial victories over the enemy.  I might add, I believe, Flight Lieutenant Beamish has actually achieved six victories as of this morning, while Pilot Officer Drummond has amassed fourteen victories as of a few days ago."

Clapping and a few 'here-here's punctuated the King's words.  He smiled and graciously waited for quiet to return.

"We are also here to honor an exceptional woman, Mrs. Charlotte Palmer.  While Mrs. Palmer, with true humility, is quite reticent to tell you this, as your Regent, I feel a duty to personally recognize the sacrifice this woman has endured in service to the Crown as well as the award I shall bestow upon her this morning.  Mrs. Palmer lost her father and her only uncle on the sacred battlefields of Northern France in the Great War.  I might add, her father, Sergeant Stanley Tamerlin, served with our Prime Minister in the Second Battalion, Grenadier Guards.  Her husband, Royal Navy Lieutenant Ian Palmer, lost his life at sea aboard HMS *Glorious* in her valiant effort to support the evacuation of Norway this last spring.  For your sacrifices to the Empire alone, we must thank you, and tell you the hearts and prayers of a grateful nation are with you."

Applause came from behind them.  The audience, however large it might be, seemed to be genuinely impressed with Charlotte's history.  Brian felt a new surge of compassion for the woman who saved his life.  She had indeed suffered grievous losses in her immediate family.

"However, we have not asked Mrs. Palmer to join us because of her sacrifices, but rather for her heroism and courage in selflessly risking her life to save an unconscious pilot whose parachute landed him in the middle of a large, relatively deep pond on her farm in Hampshire.  Alone near the pond,

Mrs. Palmer observed the aerial combat above her as well as the limp body hanging beneath a parachute.  For all she knew, the man could have been deceased already.  Without waiting for help, Mrs. Palmer dove into the lake, swam to the now wet and sinking parachute, struggled and fought to near exhaustion through the tangle of shroud lines and heavy, water-soaked fabric to reach the pilot.  I have been told by the Emergency Services that rescuing a man under a wet parachute is a very dangerous proposition, and is most often not successful.  Several professional rescuers have lost their lives in attempting to do what Mrs. Palmer accomplished.  I can only imagine the thoughts that ran through her mind during those desperate moments, but certainly she asked herself why she was risking her life for a pilot who might already be dead.

"Mrs. Palmer managed to extricate the pilot from beneath the sinking parachute.  Without knowledge of the parachute harness release mechanism, she had to drag the pilot and his parachute to the nearest shoreline.  His head and chest on the muddy edge, she struggled against her near complete fatigue to return life to his apparent lifeless body.  Miraculously, Mrs. Palmer's instincts told her to breathe life into his motionless lungs.  In those critical moments, this heroic woman managed to save the life of one of our most successful fighter pilots, none other than Pilot Officer Brian Drummond, also standing before us.  For her selfless courage, her extraordinary efforts on behalf of a helpless warrior and her ingenuity in returning life to his lifeless body, I proudly award, on behalf of a very grateful nation, the George Cross to Mrs. Charlotte Palmer."

Again, applause flooded the air, but this time the King held up his hand.  The gesture had immediate effect.

"If you will be so kind to indulge me, I would like to tell you about the George Cross."  He motioned for one of the officers to give him the medal.  The King held up a silver, straight cross with a solid dark blue ribbon with a blue ribbon bow of the same color.  "This is a new award for gallantry.  I have ordained that this Cross shall be second only to the Victoria Cross as the highest award this Kingdom can bestow.  It ranks above all orders of chivalry.  I just recently issued the Royal Warrant originating the George Cross to recognize the heroism of civilians, and on occasion the military, where no other appropriate award exists.  Mrs. Palmer is the first recipient of the George Cross.  She was enthusiastically recommended for this award by our Prime Minister, the Right Honorable Winston Churchill.  Thank you, all, for attending this brief ceremony to recognize these heroes."

The applause was loud, sincere and prolonged as the King moved to directly face Charlotte.  Several officers assisted the King in clasping the George Cross to the left shoulder of her dress.  The King talked softly with her for several minutes.  Brian could not hear what was being said but knew it had to be impressive.  The colorful words of the King in describing her heroism gave Brian a chill of gratitude as well as recognition of how close he had come to the end.  Charlotte Palmer had indeed given him a second chance at life.

The King pinned the Distinguished Flying Cross on Roger Beamish's tunic and thanked him for his accomplishments and encouraged him to continue until victory was achieved.  He then moved in front of Brian.  He saluted the King smartly.  King George VI returned the salute, received the proffered medal from the officer standing beside him, and pinned the Distinguished Flying Cross on Brian's tunic above his left breast pocket and wings.  Brian saluted again.

"Mister Drummond," the King said softly, "from my heart and from an appreciative country, I thank you for your contribution to the defense of this country and freedom itself.  As an American, I know quite well you have no obligation to be here.  As a pure volunteer from a country beyond the Commonwealth, your efforts take on even greater meaning and value."

"Thank you, sir . . . er . . . I mean . . . your Majesty," Brian stammered, catching his broach of protocol.

"That is quite all right," the King said with a smile.  "I have never been much on etiquette, and it is I who should be respectful of your courage."

"Thank you, your Majesty."

"I have been informed by Sir Quintin Brand that your flight instructor in America also served the Crown with distinction during the Great War."

"Yes, your Majesty.  Malcolm Bainbridge.  He served with Four Three Squadron and received two Distinguished Flying Crosses."

"No wonder you have accomplished so much at such a young age, Mister Drummond.  Please keep up your great work.  Keep us safe and free."

"I shall do my best, your Majesty."

The King nodded his head and smiled broadly at Brian before he turned to leave.  The audience cheered and applauded.  Squadron Leader Darling marched to a spot just in front of the microphone.  He waited for the King to depart, or for some specific signal, and then dismissed the squadron.  Brian relaxed and flexed his knees, which had grown stiff from the tension of the moment.  Darling began the flood of congratulations as the other officers moved around them.  Brian caught a few glances at Charlotte who appeared to be besieged with reporters and other well-wishers.

Air Vice-Marshal Sir Christopher Joseph Quintin Brand, KBE, DSO, MC, DFC, Air Officer Commanding-in-Chief, No.10 Group, returned, having seen the King away, to offer his special congratulations to the three awardees.

He saved Brian for last. "As a pupil of Malcolm Bainbridge, there was no question of your achievement, 'Hunter.'"

Brian was somewhat surprised the senior officer knew his call sign. "Thank you, sir."

"It is an honor to have you with us, young man." He leaned forward and lowered his voice to a near whisper. "All I ask is that you keep shooting down those frigging Germans." He stood up straight and smiled.

Brian returned the grin. "I can do that, sir."

"I am absolutely certain you can. Once again, congratulations, Brian." Air Vice-Marshal Brand offered his right hand, which was unusual for a senior officer. They exchanged a firm handshake and the silent eye communications of aviators everywhere. Good luck, they said.

All the pilots, in turn, shook hands with 'Jackstay' and 'Hunter,' and eventually with Charlotte Palmer. Even his crew, Gordon, Toldson and Jenkins, waited their turn to shake hands with the awardees. Emily Darling, the Skipper's wife, worked her way through the sea of Air Force personnel to convey her congratulations on the two men. Brian accommodated the recognition, but he really wanted to talk to Charlotte. He wondered if he would ever get a chance to talk to her.

"Mid-day meal is in one hour," announced Darling. "After that we are back on duty."

As the people began to thin, Brian inched his way toward Charlotte. When she saw his face, she excused herself from the throng of reporters. The circle of men parted for her to walk toward Brian. She flashed a please-save-me-from-this-mess smile. Brian moved toward her, and they embraced. Flash bulbs burst around them like paperbags popping in the distance. Charlotte pulled back just enough to kiss him on the cheek.

"Thank you," she whispered to him. "I have been waiting for this all week."

Brian just smiled, not knowing quite what to say. He agreed without question, as she would have known if she had felt the pounding in his chest.

"Can we go somewhere?" she asked.

"Sure." The only place Brian could think of immediately was the squadron's original Dispersal building at the eastern end of the V-shaped flight line. He led her through the crowd. Charlotte took his arm, and let him move around and past people.

"Pilot Officer Drummond," a familiar but unrecognized voice called.

Brian turned toward the call. Walking toward him also arm in arm were Group Captain John Spencer and his wife, Mary. She looked radiant, glowing, in a bright floral print dress and small white hat.  Brian felt a twinge of guilt at being caught with his hand in the cookie jar.  He fought the emotion, smiled broadly and wheeled Charlotte around slowly and gently.  Brian released his right arm and saluted the senior officer.  John returned the salute and extended his hand.

"Congratulations, my boy," he said with genuine pride.

"Thank you, sir."

"Yes, indeed," Mary said, as she leaned forward to lightly embrace Brian and kissed him on the right cheek.

Brian took several quick glances to all three sets of eyes.  Mary, true to her word, did not give him the slightest expression of additional meaning.  Then, he noticed an eye signal from John reminding him of his social duties.

"Excuse me," Brian said.  "Group Captain and Mrs. Spencer, I am honored to introduce Mrs. Charlotte Palmer.  Mrs. Palmer, the Spencers."

John extended his right hand to her, took her gloved hand and kissed it.  "We are deeply in debt to you for saving Brian's life."  Charlotte had a puzzled expression.  "Yes, of course, excuse me.  Brian's flight instructor in America was my best friend.  Malcolm, God bless his immortal soul, saved my life several times in the skies over France in the Great War, and he has given us one of our best fighter pilots," he said, nodding toward Brian.  "Malcolm Bainbridge was one of the best pilots I have ever known, so it is no wonder Brian learned so well.  It is likewise an honor to meet the first recipient of the George Cross."

"Thank you."

"Our hearts also go out to you for your losses," added Mary, as she extended her hand to Charlotte.  She turned to Brian.  "You have been extremely lucky, Brian, and we pray luck remains with you through this dreadful ordeal."

"Thank you, ma'am."

"We have some news ourselves, Brian."  He paused for a build-up of anticipation.  "Mary has given me the most glorious news."  Mary smiled.  "She is pregnant."

Brian had expected this moment.  He was fully prepared and appropriately excited.  "Congratulations to both of you," Brian said, shaking John's hand and giving Mary a kiss on the cheek.

"Yes, you must be so pleased," Charlotte added.

"Yes, we are," Mary answered.  "This is our first.  We were most fortunate since I do not have much of John's time these days."

"Mary, please."

"Not to worry, John.  I am just grateful for that moment," she said, but this time her eyes spoke much more directly to Brian, who quickly looked away, not wanting to connect with Mary.  She looked to Charlotte.  "Do you have any children?"

Charlotte lowered her eyes and chin as though a heavy weight tugged at her head.  "No.  Ian and I never were able to find the time either, I suppose."

Mary embraced Charlotte, and Brian heard her say softly.  "I am so sorry, my dear.  I did not mean to cause you pain."  Mary stood back.  "It is only by the grace of God that I do not share your pain."

John displayed obvious signs of discomfort.  "I am afraid, we must be on our way.  We managed to break away for this extraordinary ceremony, and I must return to Headquarters promptly."

"As I said," Mary offered, waving her hand widely toward the RAF facility.

John and Mary said their good-byes and departed.  Brian could not avoid the gentle sway of her posterior and the delicate movement of her shapely legs as they disappeared around the corner of a building.  Brian looked at Charlotte who was looking at him with an inquisitive expression in her eyes.  He definitely did not want to go into that area of discussion.  He smiled and nodded his head in their original direction.  As they turned to continue, another person called out his name, but this time an unfamiliar voice.

A man in a brown suit and fedora with a press label and a stern, hard expression walked toward them.  The man was shorter than Brian and about Charlotte's height.  His dark complexion gave him a serious look.

"Mister Drummond," he nodded to Brian.  "Mrs. Palmer," he nodded to Charlotte.  "Please allow me to introduce myself.  My name is Murrow, Ed Murrow.  I'm with CBS News on assignment here in England.  If you would allow me to impose for just a few moments, I would appreciate a few words." Brian nodded his head although Charlotte remained neutral, maybe in deference to Brian or resentment for the intrusion, but unable to say no.  "Just to confirm the facts," he said, pulling a small notebook and pencil from his jacket, "my notes indicate you are from Wichita, Kansas."  Brian nodded.  "This says, you left home immediately after high school, the spring before the war started, crossed the border into Canada, defying federal law, and joined the RAF."

Brian did not like the tone of Murrow's words.  He suspected a trap.  If he admitted his violation of the Neutrality Act, would the government invoke some other law and extradite him back to the United States.  Brian considered ending the conversation without answering the implicit question.  Murrow must have sensed the stiffening resistance and suspicion in Brian.

"My apologies, I didn't mean it to sound quite so coarse.  Let me explain," he said and paused to gain Brian's consent that took several moments to come.  Murrow looked around the sky above them, as if he were trying to find words in the clouds.  "The British are engaged in a desperate struggle to stop the Nazi tide.  I have been here reporting on events from the beginning. I have broadcast to America, and to the world for that matter, about the courage of the people and heroic efforts of the RAF to defend this ancient land. You represent so much that is good in America.  You are our youth, America's youth.  A young man past the age of consent, but a very young man nonetheless. The facts say you had the courage and sense of conviction to disobey a widely publicized federal law, and volunteered to fly combat missions, not in defense of your country, but in defense of freedom itself," Murrow said, pausing to smile, "a much higher calling, if I may be so bold.  You have just been awarded the Distinguished Flying Cross for skill, bravery and accomplishment in aerial combat, and by the King of England, no less.  And if that were not enough to the story, you have at your arm, a genuine heroine who has been awarded the highest civil medal for courage.  If this isn't a storybook, fairy-tale picture, I'm not sure I can imagine one.  Now, I hope I have conveyed my heart as well as my facts to you, Mister Drummond."

"Yes, sir."

Murrow waited, giving his subjects plenty of time to add to their words. When none came, he said, "Are the facts I have articulated reasonably correct?"

"Yes, sir."

"If you have no objection, I would like to broadcast your story to America and the world."

"Mister Murrow, I haven't done anything that many others around me have done.  Some have done far more than me, like 'Sailor' Malan, 'Bobby' Stanford-Tuck, even 'Cats Eyes' Cunningham, the night fighter.  For some reason, everyone wants to make a big deal of the fact that I have simply done my job, tried to do what was asked of me.  I was inspired by my flight instructor, Malcolm Bainbridge . . . ."

"Excuse me," Murrow interrupted, as he jotted notes feverishly, "you said Bainbridge, so you learned to fly once you arrived in England?"

"No, Malcolm taught me in Wichita.  He flew with the Royal Flying Corps in World War One.  He won two DFCs.  His best friend and squadron mate was Group Captain John Spencer who just left.  They helped me get here.  I couldn't have done it on my own."

"Your humility is appreciated, Brian, if I may call you, Brian.  But, no one has been with you in the cockpit of your Spitfire fighter as you faced hordes of Germans."

"He has been with me and will always be with me," Brian responded, feeling tears well in his eyes.

"Is he still in Wichita?"

The tears continued to build up, as Brian fought to control them before they ran down his face. "He died this spring in a freak aircraft accident."

"I'm terribly sorry. My sincerest, heart-felt condolences to you." Murrow waited for a head nod from Brian and sensed the struggle within Brian. He turned to Charlotte. "Mrs. Palmer, I know my thanks pale in comparison to the recognition you received today, but I offer my gratitude and the gratitude of my country for your courage in saving this hero's life."

The comment struck Brian in a very awkward way. He was not a hero. He could not bring himself to accept the reference. Brian started to object, but Charlotte took the moment.

"I feel quite the same as Brian. I simply reacted to the situation. I had no idea who he was when he landed in my pond, but I am so glad I was able to save him."

"Our nation is proud of you. It is not very often you see an award of this caliber bestowed upon anyone and especially by the King. It is an honor to have been present at such an important occasion."

"Thank you."

The urge to leave mounted within Brian. Time moved along, and he would soon be called to duty. He wanted to have at least a few moments with Charlotte.

"Pardon me for interrupting your time. I suspect the two of you wanted a private word, and I've taken enough time, but I'd still like to have your permission to tell your story. It is important for America to hear, Brian."

The young American volunteer pilot considered the possibilities. He remembered the news broadcasts listening to CBS Radio in Malcolm's living room before the war. Maybe it would be good for his parents, friends and Gertrude Bainbridge to hear. "Yes. OK."

"Thank you, Brian. I shall make every attempt to do justice to your courage as well as respect the risk in your endeavor. With that, I bid you both *adieu*. Please take care of yourself. I know we shall meet again." Murrow shook hands with both of them and walked away.

Charlotte looked into Brian's eyes with admiration and appreciation. "I had no idea you were so young and had done so much, Brian."

Brian felt something slipping through his fingers. "Age has nothing to do with anything," he said, as if he needed to rebut the statement.

"No," she said softly and demurely. "Age does not matter. Your accomplishments do. He referred to you as a hero. From what everyone has told me, I believe that description is accurate."

"Charlotte, I'm not a hero. I'm just a bloke who loves to fly. Malcolm and John were heroes. They wrote history."

"As you are doing right now."

"No," Brian objected. He turned, walked a few steps away, and then looked back at her. "Didn't you say you wanted to talk?"

"Isn't that what we are doing?"

Brian wanted to ease the tension. "We can sit down in our Dispersal building."

"As you say, then."

They walked side by side without touching this time. They were half way to the original No.609 Squadron Dispersal building when Charlotte decided to speak.

"Your age does not bother me, nor does it affect how I feel about you. It simply amplifies the significance of your achievements."

"I'm sorry I was so testy. It's just that everyone has told me I'm too young. All I know is I enjoy doing what I do, and I think I'm good at it. It scares me sometimes, although I would never admit that to my mates. This is everything I have dreamed of doing. Being a part of something great and spectacular in the air. I can remember the stories Malcolm told me years ago, and listening to the experiences of others. This is what I have dreamed of doing since as far back as I can remember. I'm not too young. I'm not too young to fly fighters in combat. I'm not too young to defend freedom. And, I'm not too young to love you."

Charlotte stopped instantly and looked into his eyes. The words had slipped out of his mouth without conscious thought. Brian recognized immediately that he had stepped a long way out on very thin ice. The distinctive snaps of cracking ice froze him, as he looked for the nearest thing to grab a hold of for support.

"Charlotte, please forgive me. I didn't mean to say that. I was just trying to say that I feel older than my years." Charlotte continued to stare into his eyes without changing her expression. She looked for something. "Charlotte?" Her expression did not change. "I don't know what to say. I'm sorry, Charlotte. I didn't mean to offend you. I seem to be doing that quite a bit these days."

This time she took his arm and pulled him along. Brian took the lead. The door to the Dispersal building was open. The inside was just as they left it, except most of the chairs had been moved to the temporary dispersal tent. Brian opened windows on either side of the building to help blow away the musty smell. A few wooden chairs remained. Brian dusted them off and

motioned for Charlotte to sit.  He looked into her eyes giving her the opportunity to begin.  She reached forward to take his Distinguished Flying Cross
in her right hand.  Brian looked down to see the ornate silver cross along with
the distinctive violet and white, wide, diagonal striped ribbon holding it to his
left breast pocket.  She turned it over to reveal a medallion attached with the
characters 'G VI R' over the number '1918' raised on it.

"Very impressive, Brian," she said.  She looked down at her award.

Brian looked closely at the George Cross.  He wanted to hold it, but
it was too close to her breast.  He did not want to risk the perception of an
intimate touch however innocent it might be.  The simple silver Cross, under
the striking solid blue ribbon, was even more impressive up close.  Charlotte
examined the medal.

"It says, 'For Gallantry' all right, just as the King said.  He told me the
bow was specific for the award to women.  I certainly do not deserve this."

"I think you do."

"Sorry.  I did not mean that the way it sounded.  I suppose I just do
not see myself as gallant, or heroic, or even courageous."

"I do."

"Thank you, Brian.  I might say whether it was fate or circumstance
that brought you to me that day, I am thankful I was there."

Brian nodded.  They looked into each other's eyes.  The light blue
tint to her eyes came out more clearly than he had noticed before.  He had
always thought of her eyes as blue-gray, but the blue came out now.  Brian
was content just looking into those magnetic pools.

"Brian," she began.  "I have so many thoughts, and some of them
are conflicting.  When I first met you, I thought you were a brash, assuming,
egocentric and arrogant young man," she said without emotion.  Brian had
never heard those words used to describe himself, and he did not like the
description.  "I have since learned my first impressions were wrong.  Although
you are apparently a master of a very violent profession, I believe you are a
gentle and sensitive man.  I must say, although it is quite embarrassing for
me to say so, you have a most unusual alluring quality to your character.  I
do not want you to read too much into my remarks, but I have grown quite
attached to you."

A warm rush washed over Brian.  He hoped her words meant more,
or at least the possibility of more.  He smiled as much to stall for his mind
to find the right words as to reflect his appreciation.  Brian exhaled as if
he had been holding his breath in anticipation.  "I can't tell you how good
that makes me feel.  I've kicked myself many times for my actions at your

farm on my first visit. I don't know how to explain my feelings, but I feel a connection with you, Charlotte. It's more than you saved my life."

"Not too fast."

"No, no. I've realized my relationship with you is far too important. I want you to control our relationship, whatever it may be, in pace, depth or . . . I don't know what. I would certainly like to know you better than I do."

"I share your desire in that respect."

Three distinct knocks on the side of the building interrupted their conversation. They both waited for a head to appear in the doorway. After a few moments when no one showed themselves, Brian rose. "Excuse me," he said, smiling with annoyance.

Brian looked out the door. Jonathan stood several yards away from the door.

"What . . . ?"

"Brian, my apologies, if I interrupted anything, but you missed lunch, and we are at Available. The Skipper sent me to fetch you."

"Sure. Sorry. Thanks, Jonathan."

"I'll wait for you."

Brian nodded his agreement, and then returned to Charlotte. "I suppose you heard that?"

"Yes. I am terribly sorry I have kept you too long."

"I'm not."

"I would like to continue this discussion when it is convenient for you."

"I'll try to call you as soon as I have some time. Can I come to see you, again?"

"Any time, Brian. Any time."

"Thanks. Now we'd better go."

As they stepped out of the Dispersal building, Brian closed the door, and then introduced Charlotte to Jonathan. The three walked back to the Operations building where the Army major waited anxiously for her return. He would transport her back to Standing Oak Farm.

Brian kissed Charlotte on the cheek and whispered to her, "Until tomorrow . . . ."

Charlotte Palmer smiled and nodded her head, and then turned to her military escort, "Shall we leave these pilots to their work, major."

"Yes, madam," he answered.

Brian and Jonathan watched her walk away, and then Jonathan watched Brian. The American could only think about what might be and hoped for much more. Jonathan waited as long as his patience would allow.

"The Skipper is waiting, Brian."

The words brought him back from his near dream state.  "Sure, we'd better go."

By the time the two reached the dispersal tent, the squadron reached Readiness status.  Brian did not want to remove his new award from his tunic just yet, so he removed his jacket and left it with Corporal Warren.  The pilots joked about the awards and the ceremony.  No one mentioned Charlotte Palmer, much to Brian's quiet relief.  The scramble call came a few minutes later.  They launched back into the aerial war over the southern coastal region of England.  The first engagement took all their fuel as the German fighter escort came in waves.  A quick turnaround at RAF Warmwell got them back in the air for a second sortie that took them into the early evening.  By the time they finished for the day, fatigue returned, and Brian's apprehension, not to appear too eager, kept him from calling Charlotte.  Another day passed.

———

# Chapter 10

They say that this lot is bitterest:
to recognize the good
but by necessity to be barred from it.

-- Pindar

*Friday, 30.August.1940*
*Headquarters, Fighter Command*
*Bentley Priory*
*Stanmore, Middlesex, England*
*10:00 hours*

### Week 8

The weather outside was near perfect, and the weather forecasters reported fine conditions over virtually the entire expanse of Great Britain. Group Captain John Spencer instinctively knew this was going to be a very busy day for Fighter Command. He took a few minutes as he sat at his desk, now piled with papers, to savor his morning tea. Yesterday's travel to and from RAF Middle Wallop, as well as the extremely rare dinner with his wife and night of pleasure in her company, brought a unique satisfaction to John Spencer. He loved Mary, and it hurt him knowing that the demands of the war on his time pained her so much. The day had been his first full day away from Headquarters since the Battle of Britain began nearly two months ago.

The ceremony had been simple, direct and strictly business, rather than like most awards ceremonies. As all military organizations around the world, the RAF was no different. They tried to add considerable pomp and circumstance to separate awards presentations above normal activities. With the desperate air war and mounting threat of invasion upon them, the more direct approach seemed appropriate. A smile washed across his face as the image came to him of King George VI, taking the time and effort to conduct the presentation. His presence and participation certainly made the award more special and memorable.

The image of Mrs. Charlotte Palmer, now recipient of Britain's highest civilian award for heroism, the new George Cross, presented an intriguing set of questions for John. He sensed as he had watched Brian walking with her that something more could be between them. She was a very attractive woman in a plain, strong, confident way, quite unlike most women he knew. She reminded him of Mary in some ways. They were about the same age, if his guess was correct. John wondered whether the connection between the two, if there was one, was an artifact of their circumstances or something else. John chuckled to himself with the realization that his young American protégé seemed to have a manner with women.  He could only guess, but Charlotte had to be ten or more years older than Brian.  Of course, in his uniform and with his accomplishments in the air, Brian seemed to be much older than he really was.  Maybe Charlotte Palmer thought Brian was older than he was?

Enough curiosity and drivel, John told himself as he finished his tea and decided to make the trek underground to the Operations room.  As he entered the gallery above the large, plotting board, map of the United Kingdom, he saw the intense gaze of his boss, Air Chief Marshal Sir Hugh Dowding, Air Officer Commanding-in-Chief, Fighter Command, and the Chief of Operations, Air Marshal Geoffrey Leonard.  John stopped to survey the map board.  Numerous h blocks, designating hostile raids, gathered across the entire coastline of France, Belgium and Holland.  He counted thirteen separate raids.  If his addition was correct, there were approximately 1,500 enemy aircraft represented by those h blocks.  No wonder the senior officers' attention was focused on the map.

John Spencer worked his way across the observation balcony keeping one eye on the board, and another on the chairs and other obstacles to be navigated.  The muffled words of the telephone talkers, the light scraping sounds of the blocks being pushed across the board with each update and the quiet conversation of the controllers formulating their tactical decisions made the room hum with activity.

"Looks like we are in for a busy day," John commented quietly to the two senior officers.

"It certainly does appear that way," Leonard responded.

Dowding did not react in the slightest way, which was normal for the stoic leader of fighter forces.  John could only imagine the internal pressure Dowding felt carrying the enormous burden of protecting their country.  The accumulating enemy raids completed their rendezvous process and began moving toward England.  F blocks began to appear on the board signifying friendly fighter units being positioned to intercept the enemy.

"Keith is committing too early," Leonard muttered to himself, referring to Air Vice-Marshal Keith Park, Air Officer Commanding-in-Chief, No.11 Group, the lynch pin of the British fighter array and covering the Southeast region.  No one responded.

They watched as the number of airborne RAF fighter squadrons grew in front of the approaching wave.  As the first squadrons engaged the enemy, Leonard saw something that caused his concern to broach his tolerance threshold.  The Operations Chief joined the senior controllers at the far end of the curved balcony.  The controller's booth looked toward the west direction on the map, directly across London.  Leonard continued to watch the board, as he conferred with his group of three duty controllers.  They pointed to the board and swept their arms across several areas, as they compared impressions, perceptions and information.

Air Marshal Leonard returned.  "Keith is concerned about meeting such large numbers with his depleted squadrons.  He is asking for support from both Ten Group and One Two Group."

"I share his concern," Dowding answered without taking his eyes off the board.  "What does he get?"

"Quintin is turning up four squadrons to cover the right flank in addition to his own commitments.  No word from Trafford, yet," Geoffrey Leonard responded, referring to Air Vice-Marshal Sir Quintin Brand, and Air Vice-Marshal Trafford Leigh-Mallory, the respective commanders of No.10 Group, to the West, and No.12 Group, covering the Midlands.

"I do hope Leigh-Mallory does not play games conserving his forces to the sacrifice of One One Group."

"Should I call him?"

"We should wait to see if he comes up."

John Spencer remained a fly on the wall listening to the grave debate. While the words did not convey the importance, John was painfully aware of the raging debate within Fighter Command espoused by Squadron Leader Douglas Bader, commanding No.242 Squadron at RAF Duxford, and promoted by his group commander, Air Vice-Marshal Leigh-Mallory, regarding the massing of fighter forces.  Bader remained a vociferous and adamant proponent of large offensive formations of several squadrons of fighters called big-wings, rather than committing single or paired squadrons.  The argument centered on time. Park believed it took too much time to form a wing, and with the relative short distances from France to England and the limited warning time, they could not afford the time to form a wing, plus it was a more unwieldy formation, less responsive to the changing battle situation.  The counter argument focused on the overwhelming odds faced by the southeast squadrons.  Most squadrons, especially in No.11 Group, were below full strength.  In some cases, there were neither pilots nor airplanes, while in most others, it was the shortage of pilots that kept aircraft on the ground.  Facing odds of five, ten or more to one, made Bader's argument far more compelling.  Dowding trusted his commanders and especially Keith Park.  He believed they would do the right thing when the time came.

John Spencer kept his opinions largely to himself, but he agreed with Park. Time, agility and responsiveness outweighed the principle of mass. No.12 Group was, for the most part, beyond the reach of German fighters, and they had the benefit of another 15-30 minutes warning time. John wanted to add his encouragement to Sir Hugh to not wait too long for an order to No.12 Group.

The map board became a confusing jumble of various F and h blocks as the battle was joined. F17, F23, F24 and F31 moved from the west. John wanted to walk over to read the controller's log to see which friendly sortie block belonged to No.609 Squadron. Three other squadrons in No.10 Group's area of operations were already airborne moving to engage enemy raids in their area. The tote board on the far wall indicated Brian Drummond's squadron was one of seven squadrons aloft out of ten total squadrons in No.10 Group. Which one of those F blocks was Brian?

The enormous air battle covering most of Southern England continued for many long minutes. The animation of the controllers to their left portrayed the seriousness of the situation. Two events on the map board changed the mood of the Operations Room. Only two squadrons had taken to the air from the No.12 Group area, and first one, then another, and then another h block was placed on the board over Northern France. A second wave of bombers and fighters was beginning to form. The picture did not look good.

The tote board showed all but two of the No.11 Group squadrons were committed. The lights stepped down status levels. Once airborne, the lights moved from blue for 15 minutes, several were in the yellow at 30 minutes and a couple were already in the red at an hour of combat. They would soon be ordered to land, if they had not already expended their ammunition or fuel. The agonizing march of the lights down the board for each committed squadron increased the intensity of the moment.

"The lads are going to be caught on the ground," Air Marshal Leonard observed with some agitation. He picked up a telephone, but then returned it to its cradle before it reached his ear. "They are going to need a covering force."

"You are correct, Geoff. Please give Air Vice-Marshal Leigh-Mallory a call. Let's get his fighters in the air while One One Group rearms."

Leonard walked to the controller's booth. More animated discussion followed. John Spencer considered joining the conversation as an observer and student. The Operations Chief lifted the telephone, waited presumably for the No.12 Group commander, and then held a protracted conversation. He turned away from the three controllers and pointed to the map board, as if the person on the other end of the telephone connection could see the same thing.

As John watched the telephone conference, Sir Hugh Dowding turned to his Staff Secretary. "What do you think, John?"

Group Captain Spencer refreshed his image of the situation map. "It looks like the Germans are getting serious," he answered rather casually.

"Yes, quite, but you seemed to be rather intent on Air Marshal Leonard's telephone call. What is your opinion about Bader's Big-Wing tactic?"

"Do you want my personal opinion, or my professional opinion, sir?"

"Are they different?"

"Yes."

"Then your personal opinion."

John thought about his choice of words. "Bader has some valid points, but I am afraid I side with Air Vice-Marshal Park. Time is more critical than mass given our circumstances. It takes too much time to form three or four squadrons into a cohesive unit, and once formed the wing is not sufficiently agile to make it practical in our conditions."

"What about the contention regarding the principle of mass?"

"We are shooting down more of them than they are of us."

"Yes, but John, their loss rate as a percentage of their total force is substantially less. As you recall my argument toward the end of the Battle of France, the rate of our wastage is the key. At this rate, we only have another few weeks before we no longer field a viable force. The German invasion window is greater than a few weeks."

"Yes, sir," John Spencer answered with solemnity.

"The enemy is destroying nearly as many aircraft on the ground as in the air, and the pilot situation is critical. If something does not change soon, you and I may be faced with taking to the cockpit again."

"Fine by me, sir. I certainly do not feel as though I am contributing to winning this battle. I would feel better with a Spitfire around me and guns in my hands."

"John, please do not ever sell yourself short. I fully realize that paperwork is not your desire, love nor forte, but you have been and will continue to serve the Crown well."

Air Marshal Leonard returned with a distressed expression in his eyes. "They have hit Tangmere, Eastleigh, Brooklands, Manston, Biggin Hill and Croydon rather hard. Several squadrons diverted to Middle Wallop. Some of the lads ran out of petrol before they made it. Park is beside himself over the slow response from One Two Group. Leigh-Mallory believes he is doing everything he can to support the Southeast."

Dowding glanced over to John Spencer with words in his eyes. "It is the frustration of the defense and desperation of our losses that is adding to the tension."

"That and the bombing," interjected Leonard. "Damn Germans are seriously damaging our aerodromes faster than we can repair them. For the last month, we have been losing pilots faster than we can supply new, green pilots. We have diverted all the pilots who have the skills from Bomber Command and Coastal Command as well as the Fleet Air Arm. We are very thin, Sir Hugh."

They could all sense the foreboding presence of the finale. Dowding decided to sit down for the first time. The other two officers joined him. All eyes turned to the map board. Sir Hugh Dowding slumped slightly at the shoulders with the future of Great Britain, the British Empire and maybe even freedom itself bearing down on him. John Spencer felt a great deal of empathy for the father of the modern British air defense system.

They watched the second wave move north as the first wave retreated. Damage reports as well as preliminary loss numbers began to trickle into Headquarters. Nearly two thirds of the No.11 Group and No.10 Group fighters were on the ground refueling and rearming. The majority of the remaining squadrons in the three southern groups were in the air. Only No.13 Group covering Scotland and the North had any significant number of uncommitted squadrons, and most of those were decimated squadrons from No.11 Group transferred north for recuperation, refit and recovery. The picture before them did not look good. John felt the urge to lash out at something; anything would do. The impotence of sitting in an underground bunker when there were empty fighter seats and a cataclysmic air battle in progress added substantially to his personal frustration and anger.

"Maybe it is time for me to jump in the cockpit and have at the bastards," John said aloud without meaning to communicate.

Leonard laughed and even Dowding flashed a rare smile.

"When was the last time you flew a fighter?" asked Leonard with sarcasm.

"Six or so months ago."

"What, a Gladiator?"

"No, a Spit."

Leonard stopped to consider what he did not expect. "Yes, but John, when was the last time you fired guns in anger?"

"In a few flights, I could come up to speed."

Sir Hugh decided to end the repartee. "John, I admire your enthusiasm, but your air combat days are past."

The hum of the Operations Room attracted their attention. The second wave began reaching their targets, and the level of activity in the room reached a new level of animation. The image before them appeared to be a repeat of the first series except all the numbers on the blocks were larger – the sortie designators, the numbers of aircraft and the quantity of F and h blocks on the board were all larger.  While they watched the developing action, John could only think about the purity of the cockpit and the contest between adversaries. He suspected the senior officers thought his desires were bravado.  They did not understand the simplicity of his mind.

One of the wing commander, assistant controllers carried a message for Dowding and Leonard.  "Sir, we just received a report from One One Group.  Biggin Hill, Croydon, Manston and Kenley are out of service due to bomb damage.  They expect Croydon and Kenley to be back in operation within the next few hours.  Northolt was also hit hard but remains operational. Regrettably, I must inform you, Air Commodore Henderson was killed in the attack on Northolt."

"Henderson?" asked Leonard.

"Terrence Henderson," John Spencer interjected.  "He was the senior controller at Uxbridge."

"What was he doing at Northolt?" asked Dowding.  The senior controller of No.11 Group was rarely away from the operations room buried underground at the group headquarters compound.

"Air Vice-Marshal Park asked him to work with the new sector controller at Northolt," answered the wing commander.

"Tragic," said Dowding.

"Just what Keith needs right now."

"Anything else?" asked Dowding.  The controller shook his head, no. "Thank you."

As the wing commander returned to his post at the far end of the balcony, Dowding and Leonard turned back to the map board.  The blocks were mixed up all over the south.  The battle was again joined.  John Spencer wanted more than ever to be in the air among the confusing mass of fighters. He looked at the tote board.  No.609 Squadron was back in the air.  Pilot Officer Brian Drummond was in his element.

"Geoff," said Dowding, "I think we must take the last of our options."

"Eliminate the rest periods?"

"Yes."

"Sir," interjected John, "the pilots are barely functional as it is.  Fatigue and exhaustion are causing as many casualties as the enemy.  They are at their limit.  Some are beyond."

Dowding nodded his head.  "John, we are all painfully aware of the demands.  I should remind you we have no other pilots."

"We have three squadrons that have been so depleted they are no longer operational," added Leonard.  "Eight more that are barely functional."

Dowding took a deep breath, looked at both men and turned to John. "We are also very near the conditions the Germans need to attempt an invasion. The only obstacle remaining is air supremacy over the landing beaches. If we lose the air battle or withdraw our air defenses, I believe, as others do, that the invasion would be inevitable. We have no choice."

"They will only be able to sustain continuous operations for a very limited period of time, and then exhaustion will accomplish what the enemy could not."

"You . . . ," Leonard started to say but stopped when Dowding raised his hand.

"I am afraid our fate is now in the youthful vigor of our pilots, my so called chicks. It is now when I hope all of us as well as the nation and the world see why I have devoted the last decade, despite the concerted resistance, to giving them the best possible equipment, support and protection. The truth is now." Dowding paused to search John's eyes. The veteran and ace fighter pilot fully understood and appreciated his commander's view. Dowding turned to Leonard. "We must reaffirm our earlier decisions. We can no longer rotate squadrons. The assault on the southeast aerodromes has been relentless, effective and debilitating. We should also prepare to move all but a skeletal force south from One Three and One Two Groups."

"Leigh-Mallory will have a conniption," said Leonard.

"And, he will also do as he is commanded."

"One Three Group is already below minimum levels for an adequate defense of the north."

"If we must abandon the north leaving only a token force, then so be it. All indications are the invasion will come across the Channel and most probably on the southeast beaches. The critical battle is here, and it is now. This is Agincourt. We must see this to the end."

"'We few, we happy few, we band of brothers.'"

"Indeed."

John Spencer found solace in Dowding's reference to the great battle of 1415 and his remembrance of the most poignant line from Shakespeare's commemorative play. He shuddered with his memory of the fatigue he experienced after protracted air battles over France at much slower speeds, less g loading and fewer sorties. The stories filtering back to the Headquarters about the exploits of the pilots in the face of such dominating fatigue, danger and odds brought a mystical quality to the situation. Sir Hugh Dowding was precisely correct – the fate of the United Kingdom lay in the hands of The Few.

*Friday, 30.August.1940*
*SS Duchess of Richmond*
*Canada Branch Dock No.2*
*Liverpool, Merseyside, England*
*18:00 hours*

The personnel assigned to the British Technical and Scientific Mission to the United States of America, less Tizard and Pearce, who were already in the United States, gathered on the dockside railing on the Observation Deck amidships – the topmost complete deck on the SS *Duchess of Richmond*. Numerous officers and sailors of the Royal Navy filled the remainder of the railings space to observe their merchant marine cousins perform the sailing tasks common to all ships leaving port. They watched, in silent fascination, the process of casting off the mooring lines that had secured them portside to the dock. For many of the team, this was their first experience aboard a ship of any type. Two, comparatively small, harbor tugs that had attached themselves to the starboard side began slowly pulling the modest size, Canadian Pacific Lines, passenger ship away from the pier and into the channel. The relatively tight confines of the dockyards made unilateral ship handling quite difficult. This was the business of the harbor pilot and tugs – a challenge they readily took up several times each day, seven days a week, to keep the Liverpool docks moving people and vital supplies into the country. Occasionally, they carried people and cargo away from England, as was the case this evening.

Maneuvering the ship into the main shipping channel took less than an hour. The observers watched the harbor pilot board the forward tug, and then their attach lines were released, freeing them for the next assignment. As the *Richmond* made way out the main channel, a freighter passed them entering port, possibly to take the dock position they had just vacated.

The *Richmond* slowly moved several miles off shore among other ships of many types in what was called Liverpool Roads – the anchorage area that supported the port. To the surprise of most of the civilians, the ship dropped anchor. A nearby Royal Navy lieutenant explained the common practice was for ships to gather in the Roads, forming a convoy, and then make their way to the opening ocean just after dawn to give the escort warships and the aircrews of Coastal Command the best chance to deal with any loitering U-boats that might be waiting to ambush the unarmed merchant ships, leaving or arriving at the ports of Liverpool or Belfast. The naval officer expected the Richmond to be at the center of a twenty-ship, fast convoy, and he understood they would have a rare squadron of Royal Navy destroyers to protect them to the mid-Atlantic, where presumably, a similar number of U.S. Navy destroyers would pick up their escort duty. Many of the civilian scientists and engineers marveled at the complexity of these choreographed shipping operations.

After their evening meal, many returned to the Observation Deck. It was not quite dusk in the shifted daylight of continuous British Double Summer Time hours implemented for the duration of the war. The setting sun illuminated the landmark buildings of Liverpool and most of the exterior portions of the city's dockyards remained clearly visible. Above the persistent hum of the ship, several of the men detected the all-too-familiar faint pulsating beat of German bombers. The north-northwest, steady, 10-knot wind gave the observers a near broadside view of the city. The thump of anti-aircraft batteries firing at the intruders punctuated the moment. They could see the shells explode at altitude. The incoming German bombing raid appeared to be 20 or so aircraft of various types. Waterspouts rose in the vicinity of the dockyards. Then, the German bombs found their targets. The flash of detonation and the rapidly emanating, shockwave condensation bands told them exactly what they were witnessing. The whump-whump-whump explosion sounds reaching them several seconds later. Many of the team members felt an odd admixture of emotions – anger with the German attack, gratitude they were still not tied up in that dockyard, nausea as witnesses of the destruction, and anxiety they were far closer to the air raid than they wanted to be. Bombs began landing in the city beyond the docks.

The Tizard Mission team would have a fitful night aboard ship in Liverpool Roads before they got underway for the New World and the former British colonies. They also feared the U-boat threat that shrouded their Atlantic Ocean transit, although they understood the speed of their ship and the convoy in which she would run gave them the maximum probability of surviving the journey without harm. The night's close-call added substantial poignancy to their mission. The high expectations of His Majesty's Government rested upon their worthy shoulders

—

*Saturday, 31.August.1940*
*RAF Middle Wallop*
*Middle Wallop, Hampshire, England*

Pilot Officer Brian Drummond taxied his Spitfire to his designated parking spot. The holes in his left wing and two jammed or damaged guns in the left wing impressed him, as he sat in the cockpit still strapped into the seat. The sweat soaked uniform offered some evaporative cooling on the warm day. Two long sorties by lunchtime left him with a pronounced headache and a growling, gurgling stomach.

"Managed to get yourself in a tangle, did you now, Mister Drummond?" asked Leading Aircraftman Bernard Gordon, as he stood beside Brian on the left wing and looked over at the damage.

"You could say that, Bernie," he said without the energy to offer a flippant remark.

"Toldson and Jenkins are on it. I shall have the engineers assess the damage. We'll try to have her repaired before the next go."

"You best step lively then. Chances are we'll be scrambled soon."

"Yes, sir," Gordon said on his way toward the ground.

Brian remained in the cockpit allowing all his muscles to go slack. He closed his eyes as he laid his head back against the seat and listened to the activity around him. The ground crews shouted commands among themselves. The heavy metallic sounds and discernible shudder in his 'PR-F' Spitfire indicated the removal of the damaged guns. He had taken a serious hit from several explosive, German, 20mm cannon shells. The horrific bang of the detonation had shaken his concentration as he broke off his attack to assess whether his machine would continue to fly and fight. The hammer raps and creaking aluminum marked the crews efforts to repair the wing.

He opened his eyes, pulled off his helmet with the oxygen mask and goggles attached, and removed his gloves. The characteristic snap of his harness release set him free from the machine. Brian slowly extracted himself from the cockpit. He groaned as he stood on the wing and stepped off onto the grass.

"Outboard two gun compartments are messed up rather badly," Gordon reported.

"Can you make her flyable?"

"Depends on how much time you give us. Maybe an hour, I should think. I don't think we can get the guns back in, until at least tonight." He paused for a reaction. Brian could only stare at the crew chief waiting for the rest of the story. "Do you want to go without them?"

Six guns would be better than none. No matter how they ballasted the aircraft for the missing guns, he would have larger lateral trim adjustments to make. "Can you give me some ballast equivalent to the guns plus half an ammo load?"

"We'll have to calculate the weight, but we should be able to manage it."

"Then, I'll go with six."

"I'll get Jenkins on it straight away."

"Thanks, Bernie. You guys are doing a great job. Just keep her flying."

"Can do, sir."

Brian ambled over to the inanimate collection of pilots sitting and lying near the entrance to the dispersal tent.  All the chairs were taken and he was too tired to go inside to find another chair.  Jonathan Kensington lay in the grass with his eyes closed.  Brian almost fell beside his friend.

"Why are we here?" asked Brian with food and lunch on his mind.

"War on, old chum, plus we are waiting for more word about Biggin Hill."

"What's up?"

"Word has it, Gerry nearly bombed the place into oblivion this morning."

"Seems to be the norm, don't you think?" said Brian, as he recalled the bombing of RAF Middle Wallop and the stories about Croydon, Hornchurch, Kenley and Manston.  The pilots from No.11 Group talked about the challenge of returning from combat and trying to find a place to land their aircraft without ripping off the landing gear or disappearing into an enormous crater.  Jonathan either had not responded, or Brian did not hear his words.  "So, where's lunch?"

"The bastards told us the lunch hour passed."

"No lunch?"

"Righty-ho."

"Isn't that a friggin' good deal?"

"Spot on."

"I'm not sure what feels more pain – my joints, my head or my stomach."

"Are you ill?"

"No, just hungry."

"The Skipper asked Madam Darling if she would prepare us some cheese and crisps."

"Excellent.  I hope she gets here before we scramble."

"Indeed."

"Of course, I may not be able to go."

"Why?"

"Took some cannon hits in my left wing.  I don't think it's anything serious other than it took out two guns and put a big hole in the wing."

"Nothing, serious," Jonathan protested.

"Bernie says he should have a makeshift repair in an hour or so, but I'll be missing the two guns."

"Are you going to go with six?"

"Sure, why not?  I asked him to ballast for lateral trim."

"You are certifiable."

"Naw, just committed."

"You should be committed."

Pilot Officer Roland 'Boxer' Stockard rose from his slumber just enough to comment. "We all agree."

"Bugger off," Brian joked.

"My, oh, my, our young Yank has truly been inculcated with the British vernacular."

Mrs. Emily Darling appeared from the south corner of the dispersal tent carrying a large tray with a white cloth covering a mound of something. "Would one of you lads be so kind to fetch the urn of tea for me?"

Johnson and Mansek both jumped to their feet and departed on their assignment. Squadron Leader Darling brought a small table outside for the tray and tea urn. Pilots began to show signs of life, as they rose from their chairs or grass to retrieve cups, mugs and glasses. Johnson returned carrying a large silver urn, Mansek, as his escort, carried a small tray of milk, sugar and several spoons.

The conversation with Mrs. Darling among them remained sparse, nondescript and light. Everyone knew it was bad form to discuss the trauma of aerial combat with women around and especially the wife of their leader. A mixture of several cheeses sliced along with bread, crackers and some fruit disappeared quickly as the hungry pilots devoured anything edible. Corporal Warren had been able to eat some lunch during their last sortie, as did Emily Darling, but both women snatched a piece of cheese from the tray before it disappeared. The tea was likewise perfectly prepared and managed to seal the enjoyment of the light but sufficient meal. Each pilot conveyed his gratitude to Emily Darling for her efforts on their behalf. The RAF could not manage to feed them, but the wife of their leader could. Squadron Leader Darling carried the tray and urn back across the carriageway, as he accompanied his wife back to their bungalow.

Brian and Jonathan lay back down on the grass in the shade of a tree on the north side of the tent. The sounds of crinkling aluminum, metallic hammering and human effort along the fighter flight line defined the ground crews feverishly working to return twelve fighters to full readiness. Faint words among the metallic sounds delivered a sense of urgency, commitment, focus and will. The ground crews would not allow their collective efforts to be an impediment to the performance of the squadron in the defense of their country. Amid the atmosphere of the flight line, Brian allowed his mind to drift near unconsciousness, meandering from one thought to another.

It had been two days since he last talked with Charlotte Palmer.  It was only partly due to the demands of fighter operations and the fatigue generated by the hours in the air and proximity of other humans in distinctly angular, brutish looking single seat, monoplane fighters trying to do him physical harm. He felt her nearness.  He sensed progress in their relationship whatever it was going to be, and he did not want any more risk that might jeopardize that progress.  Eagerness, as Anne Booth had told him, was the curse of youth. Brian wanted to be as mature and sophisticated as Anne had taught him.

As his mind began to let go of consciousness in the coolness of the summer shade, the distinct double buzz bell of the telephone brought him back although he kept his eyes closed in protest.  He yearned for the sanctuary of sleep no matter how uncomfortable the ground might be.

"Scramble the squadron," shouted Corporal Jennifer Warren, producing the usual response among the pilots.

As Brian approached his 'PR-F' Spitfire, Leading Aircraftman Bernard Gordon and four other crewmen continued to work on his left wing damage. Brian jumped up onto the left wing, but did not place his flying gear in the cockpit as he usually did.  Bernie Gordon looked from underneath the leading edge of the wing.  "Are we going to make it?" asked Brian.

"Nearly done, sir.  Well, at least the best we are going to do until tonight."

"Will she fly?"

"She'll fly all right, Mister Drummond.  I am not sure how she'll hold up when you start yanking her around the bloomin' sky, though."

The other engines began starting.  Brian stepped into the cockpit, sat down, strapped on his parachute, fastened his seat harness and pulled on his leather helmet still wet from the morning's sweat.  The seductive aroma of high-octane gasoline surrounded the cockpit.  He looked just in front of the windscreen at the fuel filler cover.  The two large fuel tanks sat directly between the engine and the cockpit.  The stain of evaporating fuel marked both sides of the forward cowling.

As Brian stepped through his pre-start checks from memory, Gordon appeared beside him to check his straps and connections, and closed the small access door.  "You should give her a couple of good speed runs and turns before you jump into a fight, Mister Drummond.  The patch should hold, but I would not want you in the middle of an engagement, if it decides to come off."

"OK," Brian said, as he reached for the start button.  "Let's get her turning over," he added, as he noticed several of the aircraft had already begun to taxi, and he was the last one to start.

Brian managed to catch up to the squadron before they took to the air.  He kept checking the large, dull silver, square patch on the wing as they reached for altitude.  A little light maneuvering in the formation told him the aircraft felt normal, except for the left bias to the stick position to compensate for less weight in the left wing.  They leveled off at their assigned altitude of 22,000 feet (Angels Two Two), as they headed southeast to their intercept point.

"Sorbo Green Leader, this is Green Three calling.  I need a few dives to check this patch."

"Looks like we have a few minutes, give her a good eval, 'Hunter,'"

"Roger."

Brian scanned the sky around them.  It was a crystal clear sky.  Perfect flying conditions.  No contrails or other specks in the sky gave him the opportunity he needed.  Brian rolled gently to his left, away from the squadron formation.

The 'PR-F' Spitfire Mark IA accelerated rapidly past 300 mph.  The thin white needle on the airspeed indicator moved across the black face and white numbers toward the small redline at 362 mph.  Brian kept the aircraft in a wings level straight dive and pulled back on the stick to stop the needle just short of 360 mph.  As the nose rose above the distinct line of the horizon, Brian glanced out at the silver patch.  Everything was OK, so far.  He kept close to the squadron as he performed another straight dive past redline to nearly 400 mph.  Still no problem.  A few quick turns with progressively harder pulls until his collapsing tunnel vision told him the aircraft was near its limit.  Brian scanned the sky around him several more times to keep himself close to the squadron, as well as make sure no one was trying to come up from behind him.

The aluminum patch did not look smooth any more.  It looked like it had a diagonal wave in it.  The aircraft still handled properly, so there was no discernible effect on handling qualities, but the change in appearance did not look right.  Brian considered whether he should return to the airfield, but also knew his friends were headed into a fight with substantial odds against them.  They needed every gun they could take to the party.  He decided to stress the patch some more to see what it would do.  He dove the machine to redline, and then pulled up and rolled the aircraft just as he might in a fight.

A loud bang shook the aircraft just like he had been hit.  Brian instantly looked behind him fighting the powerful vibrations.  He rolled to look underneath him along with a couple of clearing turns to search the entire sky.  Nothing.  Brian looked to his left wing.  The patch was gone along with additional metal from the wing skin.  The vibration remained strong but steady.  The noise of a whistling roar deafened the cockpit.

"Sorbo Leader, this is Sorbo Green Three calling. My patch came off. I must return to base."

"Roger, Green Three. Do you need any cover or assistance?"

"Negative, Leader. I can still handle her at low speed, but I'm out of the action. Sorry, Skipper."

"See you back home."

"Bandy, this is Sorbo Green Three calling."

"Sorbo Green Three, this is Bandy. We heard you. You should be clear to home."

"Roger, Bandy."

As Brian descended at slow speed, he decided to evaluate his landing configuration to make sure there was no other damage he could not see that might affect his ability to land the aircraft. He lowered the landing gear, maneuvered the aircraft gently, and then lowered the flaps and maneuvered some more to complete his assessment. At slower speeds, the vibration and whistle were substantially less than they were at high speed.

Brian landed his damaged aircraft at RAF Middle Wallop. He thought about parking the aircraft in front of their permanent Dispersal building since the crew would undoubtedly have to tow the aircraft into one of the maintenance hangars. The No.609 Squadron hangar, Hangar No.5, was still totally unusable from the bombing a month ago. Repair on the hangar progressed although very slowly. He realized it was Leading Aircraftman Bernard Gordon's decision along with the maintenance chief. He taxied back to their dispersal corner.

"Sorry the patch did not work, Mister Drummond," said Gordon.

"No sweat, Bernie. I gave her a good run through. It came off during my last rolling pull-out."

"We shall take her up to the Maintenance and Repair Section. I talked with them while you were airborne. They think we will have to pull the wing off and give you a new one."

"How long will that take?"

"A day plus or minus a little, according to the chief."

"You guys gave it a great try. Thanks for the extra effort."

"Our pleasure, sir. Anything to keep an ace in the air."

Brian smiled. He liked the recognition of his accomplishment from his crew chief, but oddly he did not like it from anyone else. He wanted to do so much more. It was certainly quite clear the Germans were not going to let the situation sit. More engagements lay ahead. Brian watched his crew position the tow tractor and tail wheel dolly behind his aircraft. He helped the men lift the tail wheel onto the dolly, and then watched them tow the Spitfire up the hill toward the hangars.

Corporal Jennifer Warren stood at the entrance to the tent. "The repair did not work, I see," said Warren.

"No, unfortunately. The guys were headed into a 50+ raid, which probably meant 30 or more fighters. They needed me in there with them."

"Yes, sir."

"Did you call Sector Control to report me out of action?"

"As soon as I saw your letters, sir."

"Good."

"Why don't you go on up to the Mess, Mister Drummond? Station does not have any spares. Maybe tomorrow or the next day."

"Bernie says they might have my bird ready in a day or so."

"There isn't much you can do here without an aeroplane."

"I should wait for the squadron to return."

Brian placed his flight equipment on his assigned peg beside the small, field desk Corporal Warren used, and then returned to sit in one of the more comfortable chairs. He motioned for Warren to sit with him. "You might as well enjoy the great afternoon weather," he said.

"Don't mind if I do, sir."

"Are things always this peaceful when we aren't around?"

"Except when the Germans bomb us." They both laughed. "I didn't get a chance to congratulate you for your award, sir. We are all proud of you."

"Thank you, Jennifer, but I was just doing my job. It is Mrs. Palmer who deserves the honor."

"Yes, sir. It was most impressive listening to the lads tell the story, and then listening to the King himself acknowledge what she did."

"She gave me a new life," Brian said, thinking of Charlotte. He wanted to be with her, to hear her voice, to feel her presence.

"She is a very attractive woman. Most regrettable she has sacrificed so much, her parents, her husband, very tragic."

"It sure is."

"You seem to have hit it off with her," Warren said rather casually.

Brian felt a flush of embarrassment, as if she could read his mind. Were his feelings for Charlotte Palmer that obvious to everyone? The telephone rang, and Jennifer leapt from her chair. Brian turned to watch her face as she listened on the telephone. She returned to her chair.

"They are on the ground at Tangmere for a quick turnaround. Apparently, there is quite the row to the east."

The information stimulated Brian's resentment for his damaged aircraft. He thought about walking up to the Operations building or even the Sector Control bunker on the north side of the base, but elected to remain with Corporal Warren. Brian could see the mission for his squadron like all the others for the last few months. He belonged with them, but he was a rider without a horse.

It was another two hours before the squadron returned. Brian counted the aircraft. Two more were missing. He checked each letter. 'PR-P,' Mansek, and 'PR-R,' Koenig, did not return. His imagination told him Koenig would probably never return. He could not see a picture for Mansek. The grim expressions added certainty to his imagination. A few grumbled comments confirmed the picture. Pilot Officer Stanley 'Slim' Koenig did not escape his burning fighter before it exploded. Pilot Officer Kormer Mansek at least managed to bail out over the Channel near Worthing. He waved under his parachute and was probably all right, although 'Boxer' Stockard believed he might be burned.

On the positive side, Morrow and Davies both picked up an additional victory along with 'Fog' Johnson and 'Spike' Darling. Squadron Leader Darling now had six victories and was also an ace. He would receive his Distinguished Flying Cross soon. The loss of comrades dampened any celebration they would otherwise have enjoyed.

Another hour passed for the pilots to complete their debriefings and to establish the condition of the squadron, before Group released them. Darling asked Brian to be ready for duty in the morning. If there were still no aircraft ready, then he would have the day off. Brian showed his eagerness for duty, but his thoughts were of Charlotte.

After a refreshing shower, their evening meal and before joining the others for drinks in the bar, Brian called Charlotte.

"Winchester Four Three Seven Nine," she answered.

"Charlotte, this is Brian." Silence on the other end of the line did not bode well. "Are you there?"

"Yes, Brian, I am here. I was wondering if I would ever hear from you again, or maybe something happened to you."

"You could have called the Station. They would have told you we have been very busy, and I was OK."

"I learned my lessons with Ian. Duty is duty, and the wives wait."

Brian liked the sound of her implication although he knew she did not mean him. He wanted to tell her about his damaged aircraft, partly to impress her, but also to provide reality to his delay in calling.  He wisely decided to avoid the reference.

"My apologies, but we have been dreadfully busy."

"I know that Brian, that is not the problem.  Is it so hard to call me for just a few minutes in the evening after you are released."

"I didn't want to be a nuisance, and hearing your voice just makes me want to see you even more."

"When will I see you, again?"

"I may get some time off tomorrow afternoon, but I won't know until about mid-day tomorrow."

"If you get the time, will you come see me?"

"If you have no objection, I sure will."

"Then, I shall hope for tomorrow."

"Great."  He waited for any response from her.  "I'll try to see you tomorrow then."

They said good-bye, and Brian was left with the vision of Charlotte Palmer.  He tried to participate in the usual evening entertainment, as they compared experiences during the day's operations and generally talked about flying.  Brian wanted to go to her tonight, but he knew he would not be awake much longer.  Tomorrow possessed a certain element of anticipation and excitement.  He could only hope she shared his feelings.

Brian joined the others in the bar.  He received a pint of beer and went outside with most of the pilots.  The majority of discussion centered on the tales of mistakes being made due to bone-numbing fatigue.  Pilots landing with the wheels up, falling asleep before they could taxi back to their parking spots, and landing at the wrong airfields.  They all knew but did not speak it . . . they were being worn down to nearly ineffective levels.

The sounds of Blenheim fighters groaning into the air one at a time gave them their first indication something was happening.  Quiet returned to the countryside, and the pilots returned to their conversation.  The pulsating drone of the German medium bombers with their unsynchronized propellers gave them answers and more questions.

"That damn sound is so bloody annoying," said one.

They listened.  No words were spoken.  They listened for many minutes. The bombers were headed north.

"Birmingham?"

"Maybe Manchester or Liverpool."

"We shall know in the morning."

"How many?" someone asked.

"Sounds like a shit pile full," responded 'Red' Burns almost to himself.

The laughter at the analogy as well as the fading sounds allowed them to renew their conversations and drink their beer. Brian felt an irritating frustration. They were fighter pilots. The enemy was overhead, and they stood on the ground drinking beer. It just did not seem right. This was not how it was supposed to be. They belonged in the air.

—

# Chapter 11

*Voilà le commencement de la fin.*
(This is the beginning of the end.)

-- Charles-Maurice de Talleyrand

*Sunday, 1.September.1940*
*Headquarters, Fighter Command*
*Bentley Priory*
*Stanmore, Middlesex, England*
*10:00 hours*

### Week 9

The third night in a row in the small Officer's Quarters room buried in the bunker complex did not help Group Captain John Spencer's mood. The quickening beat of the crescendo foretold the approaching climax. For John and many others in His Majesty's Government, the military, and the emergency services, the end was near. They could only pray to God for a benevolent conclusion. John found himself wanting Mary's embrace, as a touchstone of confidence all would be as it should be. In the haze of mounting fatigue, his concentration seemed to escape his grasp more often allowing his mind to think of Mary; his surrogate son, Brian Drummond; and his friends still in uniform standing against the tide. He longed for just an afternoon to enjoy the shades of green covering the gently rolling English hills and to smell the fragrance of the Earth carried on a gentle summer breeze. However, the war was never very far away.

The decisions made by Air Chief Marshal Sir Hugh Dowding yesterday and turned into execution orders issued this morning made the end seem ever so much closer. The elimination of squadron rotations, pilot rest periods and tightening of the medical leave definitions started the timer that could only last another week or two at the most. Fighter pilots, especially in the Southern three groups, faced a determined adversary in large numbers, but it was the debilitating erosion of their energy and concentration by multiple sorties with mind-baffling speeds, lethality, and exposure along with overwhelming odds that took the greatest toll. The pilots needed to recover their alertness to survive the staggering air battles. John knew Dowding had no choice, but the timer bell would soon mark the end.

The knock on his door brought him back to his office on the ground floor of Headquarters. His assistant opened the door just enough to push her head through. "Sorry to disturb you, sir, but Air Chief Marshal Dowding would like to see you in his office."

"Thank you."

Probably another set of further restrictions on the pilots tying them even more tightly to their cockpits. John found his small notebook and two sharpened pencils, and made the journey across the hallway to Dowding's office. The heavy blackout curtains had been pulled back allowing cascades of light to flood his grand corner office. Seated in front of his desk was Air Vice-Marshal Keith Park, Air Officer Commanding-in-Chief, No.11 Group.

"Good afternoon, Sir Hugh, Air Vice-Marshal Park," John said, as he entered the room.

"Good afternoon, John," the two men answered simultaneously.

"It has been some time since we last saw each other," said Park.

"You have been a very busy commander," John answered.

"Indeed. How is Mary?"

John wanted to tell the two senior officers she was much happier now that she managed to get pregnant and have a child to fill the void he had left in her life. However, reality served no purpose in this instance. The group commander was only trying to show his humanity in the face of adversity, a laudable trait. "She's doing quite well, thank you, sir, despite this bleeding war."

Park laughed. Dowding remained stoic.

"I dare say, she is not alone in that respect. This damnable war has called for many sacrifices from everyone, as illuminated by the woman who won the George Cross the other day. Dear God, her father, uncle and now her husband, and then she risks her own life, some would say foolishly, to save one of our pilots."

"Wasn't he one of the American volunteers?" asked Dowding.

"Yes, sir. In fact, he is the young fellow I have been shepherding – Malcolm Bainbridge's boy genius with a stick, throttle and trigger. He is already an ace twice over with one DFC and probably a second on the way. Second or third highest scoring pilot we have. As you know, Sir Hugh, I managed to take some time, gathered up Mary, and we attended the ceremony at Middle Wallop. The King presented the awards and did exceptionally well with his speech, introducing this new valor award, the George Cross. The woman, Mrs. Palmer, is in her own right a most impressive person."

"Apparently," Dowding said. Park nodded to their leader. "John, Keith and I wanted to talk to you before the commander's meeting which is due to start shortly. As you know, Air Commodore Henderson, God rest his soul, was killed two days ago. Air Vice-Marshal Park has asked for you as his replacement."

"I am honored," John answered with a slight smile and nod to each senior officer. "But, I have never been a controller."

"John," interjected Park, "it is not the technical skills I need. It is the leadership and instincts of a fighter pilot, I must have."

"I know you have never been happy with the paperwork I have asked you to perform, but you have slaved away admirably at everything I have asked. While it is certainly not the confines of a Spitfire cockpit, it is about the closest you will probably get to the air battle. Keith needs your experience."

"Indeed."

"Again, I am honored."

"I do not want to ask you to take this assignment unless you want the job. The demands on the senior controller especially in my area are enormous and inhuman."

Group Captain John Spencer laughed at the presentation of the mountain. "It can't be worse than the Red Baron."

Park laughed, and Dowding smiled.

"In those terms," said Park, "you are probably correct."

"Nonetheless, sir, if this is an assignment where you both feel I can best serve, then I am, as I said, honored, and I will be just as committed to this task as I have been to every other assignment."

"Smashing."

"Then, so be it. I will also inform you, the Air Ministry has approved your promotion to air commodore, and the King has granted you the honor of Companion of the Most Distinguished Order of Saint Michael and Saint George. Congratulations, John. Well deserved. Our apologies for the lack of ceremony, but these are the times in which we live. I am certain you understand."

"Yes, indeed, sir, and thank you very much for the honor to serve you, sir."

Both men shook John's hand. Park slapped Spencer on the back.

"Welcome to the club," Park added.

"Now, gentlemen, the rest of the staff awaits us, and we have crucial business to tend to, I'm afraid."

John Spencer knew what was coming as he followed the two commanders into the adjoining conference room filled with the senior staff of Fighter Command as well as the commanders from each of the subordinate commands. The same light filled the high ceilinged, simple but elegant room devoid of furniture other than the large table and two-dozen chairs. The view to the south overlooking London remained spectacular. Only the array of barrage balloons spoiled the idyllic scene. Dowding sat at the head of the table with his back to the tall windows.

"I shall begin with an announcement," Dowding said. "Effective immediately, I am transferring Group Captain Spencer, now Air Commodore Spencer, our indomitable Staff Secretary, to One One Group to fill the position of senior controller. I am certain you will join me in thanking John for his exceptional service and devotion to Fighter Command, and wish him every success in his new assignment."

Congratulatory comments, remarks and several handshakes punctuated the announcement. Among the words, John's thoughts flashed to Mary and their baby. The new assignment with its unfamiliar details meant even more time away from her. He found himself saying a prayer to himself for the air battle to subside in time for their first child's birth – a new, and unusually refreshing set of thoughts.

"Now, down to business." The room fell instantly silent. "As all of you are painfully aware, the enemy has been very effective in their assault upon the air defense system. Fortunately, to date," Sir Hugh knocked on the wooden table, "the enemy's intelligence apparatus has been quite flawed. For example, they have yet to strike the Supermarine Works at Woolston." He knocked on the table again. "Our greatest threat despite the struggle to repair bomb damaged aerodromes is our pilot shortage. We have tapped every source within our grasp and still we are losing pilots faster than they can be replaced. One One Group has borne the brunt of the attacks. There are varying estimates of our viability," Dowding said, nodding to Keith Park.

"At our current rates, we have maybe a week."

Air Marshal Geoffrey Leonard interjected, "The Air Ministry reported this American Sweeny fellow claims to have several score of additional pilot volunteers enroute."

"Charles Sweeny, also known as Colonel Sweeny?" asked Air Commodore Hogan.

"Yes, I do believe so."

"MI6 believes he is a mercenary and not really a proper colonel – a rather mysterious fellow."

"Perhaps so, yet as I understand things, five of the seven American volunteer pilots we had in July were recruited by Sweeny for *Le Armée de l'Air* before they came to us, so maybe we should not look a gift horse in the mouth. He is apparently well connected in the Air Ministry, as he has also proposed forming American volunteer squadrons with the additional American volunteer pilots he claims he has enroute to us. Further, Clayton Knight's committee in New York, in conjunction with the expert assistance of Air Marshall Bishop, have been quite active and successful in attracting additional American volunteers to our cause."

Air Marshal William Avery 'Billy' Bishop, VC, CB, DSO & Bar, MC, DFC, of the Royal Canadian Air Force, served as the Canadian chief recruiter and needed little introduction in aviation circles. He served in the Royal Flying Corps during the Great War, was credited with 72 victories, and was awarded the Victoria Cross for conspicuous gallantry in aerial combat during the war.

Clayton Knight also flew with the Royal Flying Corps as an American volunteer pilot with No.206 Squadron, ending the war in a German hospital as a prisoner of war. He became an accomplished artist in New York City during the inter-war years and decided he could best serve the cause of freedom by using his friendship with Bishop, and his contacts in the burgeoning aviation community in the United States and within the wealthy clientele for his artwork for fund raising. Knight had many friends and they helped.

Colonel Charles Sweeny was a rather mystical character, often described as an American soldier of fortune, who served in the U.S. Army during the Great War, fought with the Poles against the Red Army after the war, and had offered his services in various other skirmishes during the inter-war years, including the Spanish Civil War. He chose to do his part in the fight against fascism and tyranny by advertising at airfields across the United States for opportunities to fly for the French before the armistice, and then for the British. His nephew, whose given name was also Charles and was a prominent ex-patriot businessman in London, assisted Sweeny with numerous business and political contacts in the United Kingdom. Nephew Charles also used his business contacts to find access to funding for their pilot recruitment efforts.

"Indeed. I have had numerous discussions with the Minister and Sir Cyril. We have agreed to form up Seven One Squadron later this month with the surviving Americans we have and those soon to complete training and form additional squadrons, as the Bishop, Clayton and Sweeny recruitment efforts bear fruit. We also have some pilots from the far reaches of the Empire making their way to us. Unfortunately, these new recruits are weeks and months from becoming effective."

"If you adopted the big wing approach," interjected Air Vice-Marshal Trafford Leigh-Mallory, Air Officer Commanding-in-Chief, No.12 Group, "we might decrease our losses and increase our effectiveness."

Park started to respond until Dowding held up his hand. "We are not getting into that argument now. We have already taken regrettable and extraordinary measures without sufficient effect. It is with even more profound regret

I feel we must take further steps in our attempt to stabilize the pilot situation. Air Marshal Leonard, if you will," he said, motioning to his Operations Chief.

"We have worked several days," answered Leonard, "on a Stabilization Scheme. We have divided our fighter squadrons into three categories. 'A' squadrons are those in One One Group plus the Middle Wallop and Duxford Sectors. 'B' squadrons are those remaining squadrons in One Oh and One Two Groups. The 'A' squadrons will be kept up to full strength with trained and experienced pilots. The 'B' squadrons will be allowed five experienced pilots and will be principally dedicated to training new pilots to feed the 'A' squadrons while performing air defense operations as required. 'C' squadrons will for all intents and purposes be removed from operations with a sole mission of training replacement pilots for 'B' squadrons. As you can see, the scheme offers a cascade to feed the 'A' squadrons."

"Dear God Almighty, what are we doing?" protested Leigh-Mallory.

"We are trying to survive," responded Park quite sharply.

"I am certain, Sir Hugh," said Air Vice-Marshal Richard Ernest Saul, DFC, Air Officer Commanding-in-Chief, No.13 Group, the northern most fighter group, "everyone is aware and sensitive to the demoralizing effect this stabilization scheme will have on the 'C' squadrons."

"Yes, we have," responded Dowding, nodding again to Leonard.

"One One Group has been the point of focus for the enemy aerial assault," Leonard added, as if the obvious needed to be explained. "All available intelligence, and I do mean all, indicates the enemy intends to force a cross-Channel invasion over the Southeast beaches."

"We will not be able to defend the North. What if the intelligence should be a ruse?"

"And, if it is not a deception, and therefore the actual objective of the Germans," said Leonard, "we shall find ourselves with insufficient air power to defend the beaches."

"I should remind everyone of the Prime Minister's directive of a few days ago. It is clear His Majesty's Government does not intend to concede any ground to the enemy," Dowding said.

"That does not make it the correct policy," Leigh-Mallory answered.

"I suppose you are correct. It does not," the commander of Fighter Command said. He swept his hand across the room. "Are any of you suggesting we should abandon the Southeast? Do any of you want to be the first warriors to allow foreign invaders upon our shores in more than a half millennium?" Dowding scanned the room, as did John Spencer. No one appeared to be up to the challenge. "Good. Now, if anyone has a suggestion as to how we should defend the Southeast, speak now or forever hold your peace. We are standing upon the lip of the precipice. We have a choice. We either stand our ground, or we jump. It is that simple, gentlemen. Suggestions?"

"I still believe the big wing is the answer to these massed raids," offered Leigh-Mallory.

"There are too many sites to cover and too little time to form your big wings," snapped Park.

"We now face mass raids, not dispersed small unit attacks. Mass raids deserve mass engagement."

"Just as we did in the trenches," Air Vice-Marshal Sir Quintin Brand said quite sarcastically.

The squabble and debate between the group commanders continued for several minutes as the issues, simmering for weeks, finally came into full view. It had been Squadron Leader Douglas 'Tin Legs' Bader, the heroic Hurricane pilot given special dispensation for his artificial legs and leader of No.242 Squadron, who first espoused the big wing tactical concept before the Battle of Britain began. His group commander, Leigh-Mallory, was the highest-level advocate. Air Vice-Marshal Leigh-Mallory, the face of the argument, stood alone at the highest levels of Fighter Command and the Air Ministry, which did not seem to dampen his ardor. John Spencer found a number of attractions to the big wing concept, but ultimately agreed with Air Vice-Marshal Park – it was not the right time or place for such large fighter formations.

Sir Hugh Dowding allowed the debate to continue for half an hour. The staff, except for Air Marshal Leonard, remained neutral and somewhat distant from the tactical discussion. When the same ground began to show up again, Dowding ended the debate. In the end, the Stabilization Scheme proposed by the Fighter Command Staff was approved and implemented. Sir Hugh challenged the leadership of each of the group commanders to take the extra measures to bolster morale, support No.11 Group, and train pilots as quickly as prudently possible. The conference disbanded with Air Commodore John Henry Randolph Spencer, CMG, DFC, still seated. The darkness of his mood absorbed the light of the bright day. The thoughts were unmistakable – maybe this was the beginning of the end and not the ending they all prayed for. Maybe Great Britain would be just one more victim of the indefatigable Nazi war machine?

———

*Sunday, 1.September.1940*
*Standing Oak Farm*
*Winchester, Hampshire, England*
*13:15 hours*

**D**riving away from RAF Middle Wallop in Roger Beamish's trusty Morgan down the A343 carriageway toward Salisbury with the sounds and sight of the 'PR' Spitfires climbing at full power had not given Brian Drummond a pleasant feeling. He knew he belonged with the nine fighters heading into battle. There was some solace when he heard that the Maintenance Section had decided to replace his wing. A replacement aircraft would not be delivered for a week or so, and a wing was expected from Woolston later in the afternoon. It was possible his 'PR-F' Spitfire would be ready in the morning. The repair crew intended to work through the night. Squadron Leader Darling, only with the approval from No.10 Group Headquarters at Box, Wiltshire, granted Brian a 24-hour pass. He was to report back for duty by noon tomorrow. He was the first pilot in more than a week to get a rest break, and from all the indications they had, probably the last pilot to enjoy a break for some time.

The telephone call to Charlotte Palmer heightened his anticipation. She sounded equally as excited as he was, which Brian thought was a good sign. The afternoon drive with Beamish's diminutive but agile Morgan Sportster brought a simple but satisfying pleasure to the journey from RAF Middle Wallop.

He stopped on the crest of the last hill, as he had done before, overlooking Standing Oak Farm. The view was idyllic as if out of a painting or postcard. The hills surrounding the farm buildings were not forested, but trees were spread out just enough to allow the grasses to grow well. Cows dotted the far hillside.

Brian took a deep breath and swallowed hard. He let out the clutch pedal slowly giving the engine just enough gas to keep it from stalling. The car rolled down the hill toward the farmhouse, barn and other buildings. As he approached, Charlotte Palmer stepped out of the house onto the stone porch. Her hair was pulled back into a short braid. She wore a light, floral print dress that highlighted her figure and accentuated her shapely legs. Charlotte did not need fancy clothes, make-up and jewelry to illuminate her elegance.

"I was wondering if you had a change of heart," she said, as he extracted himself from the tight driver's seat.

"What do you mean?"

"You sat atop the hill for the longest time."

"I just wanted to take in the beautiful scene," he said, not wanting to tell her he was gaining his nerve to come down and deciding how he should conduct himself.

"What do you think?"

"About wha . . . oh, the scene. I think this is one of the most beautiful places on Earth."

"Flattery will get you nowhere, Brian," she said, stepping off the porch to give him a kiss on the cheek, which he returned.

Brian wanted to give her a hug, to feel her body pressed against his, but he restrained himself. They stood side-by-side looking out across the pond.

"Not flattery, truth."

"Thank you, and I think so as well. My grandparents, my father's parents, bought this farm fifty some years ago."

"How big is the farm?"

"About 175 acres, I think."

"Better than a quarter section. That's quite a bit of land."

"Especially in England which is why I cannot leave. I must keep the farm afloat."

"I've never had a chance to really see it."

"Then, if that is how you feel, let us do better the second time around."

Brian remembered the earlier walk. "I mean, the first time I was probably more worried about you, and did not pay much attention to the land."

"Let us take a walk, then shall we?"

Charlotte and Brian walked casually, almost meandered, around the large pond. It looked even larger on the shoreline. While he could swim well enough to be comfortable in the water, he could not imagine having to face what Charlotte did a month ago. He did not want to talk about his close encounter and her rescue, but he could not escape the thoughts. They walked up the hill Charlotte had stood on when she saw him land in the pond. They walked across several adjacent hills surrounding the buildings. She pointed out the old, short, stonewall that marked the boundary of her property, and told Brian some of the history of the farm and country around them.

Her delicate, clear and confident voice added to the strength he felt around her. He never sensed any vulnerability in her. Charlotte possessed Anne's confidence and Mary's elegance from his perspective. Curiously, Brian wondered if she might also have Rosemary's ardor and energy. He enjoyed being near her, like he might like being close to a wellspring of perpetual youth and invulnerability.

They spent the remainder of the afternoon finding humor, levity and connection among a wide variety of subjects and tasks. The cows were milked at their normal time before their udders filled completely. They enjoyed a leisurely early evening meal of chicken, potatoes and green beans, all grown on the farm. The food and drink seemed trivial and distant compared to the enthralling conversation that continued unabated into the late evening.

Brian recognized the hour, did not want to leave, but knew he must. Even though he did not need to return to duty until mid-day Monday, he acknowledged to himself he could not risk overstaying his welcome. They both enjoyed the laughter and companionship. He felt the tentacles of their connection – their bonding – reaching out between them. He knew for the first time she liked being around him, and it was more than courtesy or pleasantries.

"I should be going. It's getting late, and it'll take me an hour to make my way back to Middle Wallop with these damn blackout headlamps." Charlotte rose from her chair, picked up their wine glasses and walked to the kitchen sink. She hesitated at the sink and looked out the window. Brian struggled to think of anything that could have caused the significant mood shift. "You can't see more than a few yards, which means you've got to go real slow."

She nodded her head probably more to acknowledge his words than agree with the content. Brian wanted to ask her what was wrong, but chose to let silence fill the space until she was ready to talk. Minutes passed, and he began to feel uncomfortable with the void. He stood up. The subtle creaking of the chair caused her to turn around.

Charlotte's face sparkled in the soft light, and then he realized her face was wet from tears silently flowing down her cheeks and dripping from her chin. What had he done? What had he said to cause her such pain? He wanted to reach for her, to comfort her, but remained frozen by the doubt that blanketed him.

"I'm sorry, Charlotte. I didn't mean . . . ," he stopped when she held up her hand.

She smiled faintly. "You did not do anything wrong, Brian, but I am about to, I'm afraid."

"Then, don't do it," he said without thinking.

Her smile grew although the tears continued to flow. Charlotte walked slowly toward Brian. Her face distorted by some strange combination of pain and pleasure. She reached up to place both hands just below his ears and kissed him. Her eyes closed and her lips moved against his. He reached up to her shoulders wanting to embrace her strongly but fought to slow the pace. She stopped, pulled back just a little, but did not release her gentle grasp of his head.

"Brian, I know everything I am about to say is wrong and against everything I believe in, but I cannot seem to help myself." She paused for a response. He nodded his head. "I want you to stay with me." Her tears continued to flow.

Pilot Officer Brian Drummond – an American volunteer fighter pilot with the Royal Air Force – reached up to wipe away her tears. Her skin felt like the softest silk. A blue fire burned in her eyes like he had never seen before. His heart pounded in his chest, partly from excitement of what might be and partly with apprehension.

She wrapped her arms around him pressing against him. For the first time, Brian enveloped her with his strong arms and held her tight. "Oh, Brian. Please tell me I am not wrong," she whispered.

Brian did not know what to say. He held her close for several minutes. As he felt her grasp relax, he released her.

She placed her hand on his chest. "I can feel your heart pounding. You must be as scared as I am," she said, taking his hand and placing it on the middle of her chest. He could not avoid the curvaceous, soft flesh of her breast. "My heart is pounding as well. I have not been so scared since my school days."

Brian wanted to tell her how he felt, but embarrassment kept him from speaking. He simply and gently touched her cheek, and then ever so lightly brushed his thumb across her lips to feel the delicately curved, soft surface of her full lips. "Me, too."

"Will you?"

"What?"

"Stay with me."

Brian tried to smile but found the effort too much against his apprehension. He was more worried about consequences than actions. He wanted so much more. He yearned to feel the contours of her body, to feel the heat of her flesh and joy of her touch. It was the fear of losing whatever connection he might have with her that kept him speaking and moving closer.

Her face drained of warmth just before she turned away from him. "I am dreadfully sorry. I have made such a fool of myself," she said, as she walked away from him. He wanted to speak. His mouth moved as if to form words, but no sounds came out. "Please accept my humblest apologies for my adolescent behavior. It has been so long since I have had to do this that I have misjudged the situation."

"Charlotte," Brian said forcefully and much louder than he wanted. She turned to him from several yards away, tears returned to her face. Brian surprised by his response fought to control his emotions. "Charlotte, please," he said with softness returning to his voice. "I am scared. I am very scared. When I first met you, I offended you when I kissed you in the barn. My relationship with you is very important to me, and I don't want to screw it up." She stood motionless, waiting for more explanation. Her tears stopped. "I don't know what to say, or what to do, or how to act. I don't want to make another mistake."

She smiled slightly. "Brian, we seem to both be afraid of what might happen. I am afraid you are thinking of me as an old woman, too old for you. I feel an attraction to you I am not certain I have ever felt before for any man. It has also only been a couple of months since my husband died, and I am embarrassed that I have these feelings for another man while I should still be grieving his loss."

"I understand. This war is making everything screwy. I have great respect for your loss, and despite my attraction to you, I don't want to do anything more that might offend you. I am willing to wait for our relationship, even if it is only as friends, to develop at a pace you are comfortable with."

She laughed from deep inside – a good, strong laugh. "We are like two school children not knowing what the next steps are and not wanting to admit we do not know." She walked toward him. "Since you want me to control this, I shall rise to the occasion." She grasped his hand and led him to the bedroom.

Charlotte undid his necktie and began to unbutton his shirt. She stopped and looked into his eyes. "Wait," she said although it was clear she did not expect a response or action. "Since you want me to take the lead and break the ice, I have only two requests before we go any further." Brian concentrated on her eyes and nodded his head. "It has been a long time since I have done this." Brian nodded his head again. "I want you to be patient with me."

"I can do that."

"And . . ."

"And?"

"And, I want to hear you."

"What?"

"I want to hear you. I want you to let go, release the feelings within you. I want to hear the pleasure we will enjoy."

"Will I hear you?"

"I do not know. It has been so long."

"I can be patient and slow. I can stay with you."

"I shall try, Brian, but it has been so long."

He smiled warmly. "We have all night."

Charlotte Palmer returned to her task until all his clothes littered the floor around them. She stood back from him just a step to admire his naked body. "Dear God, Father of us all, you are quite the specimen."

He smiled more broadly. "All the better to love you with, my dear."

She jumped into his arms and wrapped herself around him. Brian rocked her and swung her gently, holding her fully off the floor, as if they were dancing on a cloud. She stopped, kissed his chest, and then slowly moved her hands over every inch of his smooth, creased body. She cooed, gurgled and purred as she enjoyed the curves of his muscles, front and back. She saved the object of her interest for last – touching him, stroking him – until he felt as if he would burst.

Charlotte stood back from him again admiring the presence he held. She started to unbutton her dress, and then stopped when he shook his head, no. Brian moved to her, taking each button ever so slowly with pauses between each button to caress her shoulders, her hips, her head, her cheeks, her breasts, her lips, her buttocks, her legs, until her clothes lay on the floor over his uniform. She wriggled as though she itched all over. He completed his task and stood back to admire her full, exquisite curves. She wriggled again feeling a little embarrassed being absorbed by his eyes. She could feel his readiness. Charlotte pressed against him to communicate her awareness and her need.

Brian lifted her into his arms and carried her to the bed. He did not pull back the bedspread as he ever so gently lowered her to the bed. Charlotte looked up at him, warm with anticipation. She stretched both arms out toward him, gesturing with her fingers for him to come to her. She smiled across her entire face. She wanted him joined with her. He moved slowly to her.

They touched but did not speak – the internal heat and desire evident to both of them. Brian flirted with their union, teasing her desire, stimulating her demand. Great care went into each step in the slow dance toward their pinnacle. Charlotte reached to him several times with her arms and her legs, trying desperately to draw him to her, but Brian resisted his own urges to heighten, enhance and invigorate her ascent. Gradually, ever so slowly, he gave her just enough of himself to push her desire to even higher levels. She groaned, pleaded and ordered him to complete their union, as she wrestled with her unleashed passion and his commitment to her pleasure. He made their union take forever in the frame of passion until she was near the rise. They both groaned together when they achieved the limit of their joining.

His efforts, in time, bore the fruit he sought in rich, full, deep measure. Her body stiffened with enormous rigidity as the convulsions and panting of her ultimate pleasure shook her for what seemed like several long minutes. Charlotte reached for his shoulders and pulled him firmly to her, feeling his weight and allowing her breathing to fill his ear. She whispered softly among her labored breathing a litany of gratitude and satisfaction. He continued to move just enough to maintain their full union.

As her energy returned, they explored positions, angles, access and additional sites of pleasure, as the two lovers developed their own repertoire of physical enjoyment. Their bodies glistened with the wetness of their exertion. They slid against each other like well-oiled machine parts perfectly matched to the task. Charlotte enjoyed two more ascents to the summit before she demanded satisfaction of her second requirement. Brian positioned her gently to fulfill his commitment to her. He kissed her deeply and passionately, as his painful pleasure took him over the top. He made certain she could feel his groans deep within her chest, as their bodies became electrified and shook with his release.

Brian's descent from the peak inspired her to take the superior position. She touched him, caressed him. Tears began to stream down her face and drip onto his chest. His expression conveyed his puzzlement.

"It is nothing," she said, as her sobs became more pronounced.

"Then, why are you crying?"

"I'm not."

Brian touched her chin, as if to retrieve a teardrop on his fingertip. "Then, what are these?" he asked, holding his finger in presentation of the evidence.

"I have never felt so much pleasure, so much enjoyment of being a woman."

"I think we have both wanted this for some time. It is everything I dreamed it would be."

"Yes."

"I feel closer to you than I have ever felt to anyone. It is more than physical. It is so much more."

"Oh, Brian," Charlotte nearly wailed. "What have I done?"

He reached to her with both hands to soothe whatever pain she now felt. He did not want her to regret what they had done. He wanted her to feel as complete as he felt. Brian could not ask the questions that raced through his mind as she continued to cry.

"You are right," she cried. "I have wanted this since the first time you came to me. I have wanted you more than I have ever wanted a man. My desires bring regrets for my insufficient mourning."

"It's OK, Charlotte," Brian said, finally realizing the struggle within her. "Life must go on to honor those who have gone before us. We owe it to them."

"Oh, Brian, what have I done?" she repeated, covering her face with both her hands as though she was now embarrassed by her actions and exposure.

He pulled her down to him and rolled them side-by-side. He held her, stroking her head and hair. They remained joined until her sobbing subsided, and then dissipated. For the first time, Brian became aware of the dampness beneath them, and the chill it brought. He moved her gently until they were under the covers. They lay like spoons in a drawer, as he continued to caress her.

"Brian, I am so confused and that is not my nature, which makes my confusion all the more troubling." He did not answer or alter his efforts. "It is like I am a school girl again, and I do not know what is right, or what is wrong." Brian wanted her to talk. "I want this to be right, but it is like anything this good must be wrong. I do not want you to be reviled by my weakness."

"Charlotte," he said softly, "that is utter nonsense. What we experienced is the passion that is ours. We must hold it, together."

"I hope you are right."

"I am."

"How did you get so wise?"

"I didn't. It just comes to me," he said, as he began to tickle her ribs. Laughter and writhing rewarded his efforts to lighten the moment and distract her from the emotional struggle within her soul.

Charlotte Palmer sat up and faced him with a smile on her face. "I want more," she said.

Brian returned her smile and initiated their lover's play.

———

*Monday, 2.September.1940*
*RAF Middle Wallop*
*Middle Wallop, Hampshire, England*

By the time Brian returned to RAF Middle Wallop, No.609 Squadron had already departed for their auxiliary airfield. He found Leading Aircraftman Bernard Gordon and his crew along with numerous other maintenance personnel and two civilian engineers from the Supermarine Works at Woolston. They had struggled through the entire night removing and replacing the wing on his Spitfire. When Brian found them, their overalls were thoroughly soaked as they continued their efforts to ensure nominal alignment of the new wing along with the flight control rigging. Even the Supermarine engineers looked bedraggled. The fighter had not gone back together as easily as it had come apart.  As Brian observed the enormous effort, the physical evidence of the team's exertion and unmistakable signs of deep fatigue, he could not ignore the feelings of guilt for his pleasure the previous night while his crew labored so diligently.  No one beyond Charlotte would know his feelings.

The mid-afternoon post-maintenance check flight took the better part of an hour and found several squawks that had to be remedied before he could return the machine to duty.  Two additional short flights had been needed to bring the 'PR-F,' Vickers-Supermarine, Spitfire Mark IA back to the exquisite handling qualities he was familiar with prior to the damage.  After the last flight, he taxied the fighter to the gun pit at the northeastern corner of the airfield.  He worked with the crew to align the guns with his desired convergence at 250 feet.  Gordon, Jenkins and Toldson scrambled, as if they had freshly arrived at the flight line to refuel and rearm Brian's Spitfire.  He flew down to RAF Warmwell at high speed and low altitude to avoid any solo combat – the demise of too many over anxious, enthusiastic young fighter pilots.  The squadron was expected back at any time when he landed and checked in with the airfield operations staff.

With his aircraft safely on the ground, fuel tanks topped off, and no one around, Brian took whatever few minutes he might have to sleep.  The dull aches he felt brought back vivid images of the previous night's efforts and rewards.  They had not wasted the little time they had together sleeping.  While they might lie quiescent for moments between episodes of passion, the two lovers had managed to use the night for intimacy.  The exertions of their lovemaking had left them both physically exhausted by dawn.

The distinctive humm of Merlin engines brought him back from his dream state.  The popping protestations of idling engines on short final and the call letters on the fuselages marked the return of No.609 Squadron.

"I didn't expect to see you here," Squadron Leader Darling said, as the first pilot to approach their dispersal tent.

"The lads worked hard to get the wing back on.  I figured if I got the machine running and there was still daylight I ought to get back into the fight."

"Glad to have you back."

Before he could respond, the other pilots approached slapping him on the back or punching him in the shoulder as they teased him about taking a day off work.  They had flown their third sortie of the day, and it looked as though they might have to go back out again soon.  When they reported their refueling and rearming complete, the squadron went to Readiness alert status.

Jonathan Kensington made sure they were beyond earshot of the others before he asked his friend, "So, how did it go?"  Brian's enormous smile told the story. "You bugger. Are you going to have every woman on this bleedin' island?"

"It's not that way."

"Oh, isn't it?"

"No. This is different," he hesitated, "not that I don't really like Rosemary and all."

"Oh, stop, you twit. My sister is old enough to defend her own honor. This woman could be your bloomin' mother, for Christsake."

"She's not that old."

"Really?"

"She's only a few years older than me, but I think we finally connected last night."

"You bugger."

"Jonathan, she is special."

"All your women are special."

"Yeah, but she's different."

"What beside she saved your soddin' life, was awarded the George Cross for her heroism, has all the right parts, and is a rather attractive, older woman?"

"You have such a disgusting way of putting it."

"Any untruth, there?"

Brian wanted to argue with him just to put his observations on a different plain. "No."

"Right . . . then, there you have it. So, are you going to marry the woman to absolve her sins?"

"Jonathan," Brian protested.

"Come now, Brian. Dunk the false modesty. We are defending our country against the heathen invaders and having our way with women. Why is this any different?"

"There are a thousand reasons."

"Really?"

"Sure."

"Like?"

"Like . . . I think I love her."

"That's what you said about Anne, for God's sake."

"It's different."

"What's so bleedin' different?"

"I don't know, but it is."

"Scramble, scramble," shouted Squadron Leader Darling, interrupting their conversation.

Both pilots took off running toward the tent, grabbed their flight equipment and joined the others racing to their aircraft. They were in the air in short order climbing for altitude. They engaged a moderate sized raid headed toward Portland. Red Section took on the bombers while the remainder of the squadron kept the fighters busy. The encounter did not take long. The squadron of enemy bombers dropped their ordnance prematurely, so they could turn and run. The British fighters gave chase for a short time, and then broke off the engagement. They were directed back to RAF Middle Wallop, landing in the late afternoon.

—

*Tuesday, 3.September.1940*
*Cabinet Room*
*No.10 Downing Street*
*Whitehall, London, England*
*17:10 hours.*

The day had already been a long one for Winston Churchill. The news from the Air Ministry as well as the War Office continued to be grim. The incessant and effective assault by the Germans on the Air Defense System of Great Britain persisted despite all efforts to stem the tide. Chief of the Air Staff Air Chief Marshal Sir Cyril Newall had convinced him earlier in the day that Air Chief Marshal Sir Hugh Dowding had taken all and the last of his available actions to stop the bleeding of his pilots. The loss rates remained absolutely precipitous in every area although the valiant pilots were managing to inflict greater losses on the enemy. According to Newall, Fighter Command lost 25% of its pilots in the last two weeks alone. They needed a miracle. They needed just a little relief for Fighter Command and especially No.11 Group covering the Southeast. Minister of Aircraft Production Lord Beaverbrook had accomplished his initial task of stimulating wartime aircraft production despite the decades of inadequate and embarrassing lack of investment. They had aircraft. They did not have pilots.

Reports from the Home Guard about the bombing damage from the raid on Liverpool Saturday night added insult to injury. Fighter Command remained nearly impotent in their ability to engage night bombers. The Liverpool attack had been the largest to date with what the Air Ministry estimated to be about two-dozen medium bombers involved. The enemy's bombing accuracy had been nearly perfect from the reports Churchill received. The docks were still burning as the Fire Brigade and other emergency services fought to gain control of the blaze.

The approaching, scheduled visit from Colonel Stewart Menzies, Chief of the Secret Intelligence Service, probably would not improve his day. Photo reconnaissance from the previous day, which Churchill examined himself, revealed an undeniable expansion of the invasion force build-up all along the Channel coast despite the best efforts of Bomber Command and the Royal Navy to slow the process. There could be no other reason for the enormous quantity of barges, ships and bivouacked troops than a serious cross-Channel invasion mission – Operation SEALION, the Germans called it.

The buzzer on his interphone took him from his thoughts. "Yes."

"Colonel Menzies to see you, sir," announced his Private Secretary John Martin.

"Show him in please," Winston said, as he stood to greet his chief spy.

"Good afternoon, Prime Minister."

"It is not particularly good, Stew, but your friendly greeting is acknowledged."

"I am afraid, Prime Minister, I am not going to make it any better."

"Splendid," Churchill responded sarcastically.

Menzies opened the locked metal case manacled to his left wrist. "I have three messages you should see. The first two are ULTRA intercepts we decoded early this morning, and the third is a message I propose to release to the War Cabinet and the Defense ministries." Menzies handed the papers to the Prime Minister. The first two were typewritten on very thin, presumably rapidly destroyable paper, while the third was mimeographed on conventional paper.

---

## MOST SECRET - ULTRA

```
SECRET
DATE:  2 SEPTEMBER 1940
TO:  OKH, OKM, OKL (Service High Commands)
FROM:  OKW (Armed Forces High Command)
BREAK
OPERATION SEALION.
        OUR LEADER HAS DECIDED UPON 20TH
SEPTEMBER AS THE GLORIOUS DAY FOR EXECUTION
OF OPERATION SEALION.  OKM TO COMPLETE FINAL
PREPARATION FOR TRANSPORT BY 13TH.  SEA
INTERDICTION TO COMMENCE ON 18TH SEPTEMBER
TO GAIN AND MAINTAIN CHANNEL APPROACHES.
OKH TO COMPLETE LOGISTIC PREPARATION AND
TRAINING.  ALL REST LEAVES TO BE CONCLUDED
BY 18TH SEPTEMBER.  OKH TO USE PLAN PORPOISE
```

```
FOR INITIAL OBJECTIVES.  OKL TO CONCLUDE AIR
SUPERIORITY ACTIVITIES BY 10TH SEPTEMBER TO
PERMIT FINAL PREPARATIONS UNIMPEDED.  CO-
ORDINATION COMMUNICATIONS REMAIN UNCHANGED
FROM FINAL OKW CONFERENCE.  OUR LEADER EXPECTS
FULLEST CO-OPERATION AND BEST PERFORMANCE TO
ACHIEVE ULTIMATE VICTORY OVER ENEMY FORCES.
ALL PERSONNEL SHOULD KNOW EVERY DIPLOMATIC
EFFORT HAS BEEN MADE WITHOUT SUCCESS TO AVOID
INVASION.  PRAISE THE LEADER.
END
SECRET
```

## MOST SECRET - ULTRA

---

"Do we know what Plan Porpoise is?" asked Churchill.

"No, sir. It is one of our highest priority interests. We are also using every available source in our operations to establish their battle plan and order of battle."

"At least we know the date."

"Yes, sir."

"What is the likelihood they could execute earlier?"

"Anything is possible at this stage, Prime Minister. The likelihood would have to be considered remote. The Air Force still makes them work for every bomb they drop. As long as the Air Force can maintain control of the skies, one of the preconditions cannot be met."

"It may not be much longer."

"Pardon me," Menzies said incredulously.

"Stew, the Air Force has managed to hang on against unimaginable odds a month longer than most experts expected. All knowledgeable estimates indicate they may have another week of viable resistance, and then we shall be forced to withdraw what air forces we have to give the Home Forces a modicum of opportunity to defend our island."

"Is it that bad?"

Churchill lowered his head and cleared his throat. "I am afraid so."

"Then, you shant like the next message much."

The Prime Minister placed the top message at the bottom of the three.

---

## MOST SECRET - ULTRA

```
SECRET
DATE:  1 SEPTEMBER 1940
TO:  SECRET STATE POLICE GROUP COMMANDS
FROM:  HEADQUARTERS SECURITY SERVICE, EMPIRE
CENTRAL SECURITY OFFICE
BREAK
        FINAL PREPARATIONS IN SUPPORT OF
OPERATION SEALION AND OCCUPATION OF GREAT
BRITAIN ARE TO BE COMPLETED NO LATER THAN 15TH
SEPTEMBER.  OUR LEADER HAS DECIDED ON THE
FOLLOWING UNIT LOCATIONS.
1.  LONDON -- HQGB
2.  BRISTOL
3.  BIRMINGHAM
4.  LIVERPOOL
5.  MANCHESTER
6.  EITHER GLASGOW OR EDINBURGH
        DR FRANZ SIX HAS BEEN APPOINTED BY OUR
LEADER AS HEAD OF SS HQGB.  TARGET LISTS TO
BE SUBMITTED TO THIS HEADQUARTERS BY 12TH
SEPTEMBER.  ALL OPERATIVES MUST ABIDE BY TARGET
PRIORITIES TO GAIN MAXIMUM EFFECTIVENESS IN
SUPPORT OF OPERATION SEALION.  INFILTRATION
UNITS MUST BE AT FULL ALERT BY 18TH SEPTEMBER.
OPERATIONS EXPECTED TO COMMENCE IN ACCORDANCE
WITH OPERATIONS PLAN SEALION.  DISSIDENTS
OFFERING RESISTANCE MUST BE ELIMINATED SWIFTLY.
HAIL VICTORY.  HAIL THE LEADER.
END
SECRET
```

## MOST SECRET - ULTRA

---

"Target lists?"

"This is the Gestapo. We believe the target lists are individuals identified by the SS as leaders of the government, military and industry, and probably anyone who has a history of anti-Nazi activity."

Churchill exhaled a loud grunt. "Then, I should be at the top of their list."

"I am quite certain you are, sir."

"If the SS and Gestapo are licking their jowls, then I suppose *Herr* Hitler must be serious about this invasion business."

"Should we consider evacuation?"

"No!  Absolutely not!  I shall stand at the steps of Number Ten until there is no breath left in this old body."

"But, Prime . . ."

"No, buts, Stew.  We shall show no quarter when it comes to English land.  We shall not fold our hands like the French.  We shall stand."

"The last message," Menzies said to change the subject, "is our proposed distribution to the Armed Forces."

---

## MOST SECRET -- OFFICER ONLY

Information has been received from various reliable sources that enemy invasion preparations are near completion;

Invasion assault preparations can be expected at any time with likely landing beaches along Southeast coast;

All forces and populace should be extra vigilant for parachute or boat infiltration of Home Islands.  Any suspicious activity should be reported to Constabulary and GHQ Home Forces immediately.

---

This message must be treated as OFFICER ONLY and should not be transmitted by telephone.  Admiralty, War Office, and Air Ministry are in possession of this information.

      *H. Pritchard*
      Lieut.Colonel. G.S.

M.I.14.
0535 hrs.
2.9.40.
Distribution:
D.D.M.I.(I). (for D.M.I.).
G.H.Q. Home Forces.
G.H.Q. (Adv.) (I), Home Forces.
M.O.3.
File.

## MOST SECRET -- OFFICER ONLY

---

"That should be sufficient," said Churchill, as he exhaled deeply.

"Should we distribute this to the War Cabinet?"

"No.  I shall handle that duty."

The Prime Minister looked his Chief of the Secret Intelligence Service directly in the eye and held his eyes for several moments. "Is there anything else we can do?"

"For what?"

"To help our fighter pilots. They need a small miracle."

"We have examined all the possibilities we can think of, Prime Minister. The best we can do at the moment is collect as accurate intelligence as we can."

"Yes, yes, I know. Have you seen any hint of German Air Force response to our bombing of London last week?"

"No, sir. We are watching and listening."

"Well, I suppose we can still hope the little corporal takes the bait."

"Is there anything else, Prime Minister?"

"No. I suppose not, Stew. Have a good evening."

Menzies departed closing the door behind him and leaving the Prime Minister with his thoughts and burdens. There was little doubt in Winston's mind they had yet to see the worst of this situation. Even Winston Churchill was finding it difficult to see any brightness to the military status and readiness for what seemed inevitable. He searched his mind and brooded over the dwindling alternatives and options.

———

*Tuesday, 3.September.1940*
*The White House*
*Washington, District of Columbia*
*United States of America*
*14:30 hours*

"Excuse me, Mister President," announced Missy LeHand, the President's executive secretary, as she waited at the door to the Oval Office. She waited for President Franklin Delano Roosevelt, 32nd President of the United States of America, to complete his handwritten note on his personal stationary, probably to a friend or supporter. "Misters Stimson and Knox are here with the final destroyer agreement for your signature."

"Please, let's not keep these gentlemen from their duties," the President said, motioning for her to let the men into this office.

The President remained quite pleased with his masterstroke appointment of the two prominent Republican politicians to the key posts of Secretary of War and Secretary of the Navy just over two months earlier. Both men had immediately risen to the challenges facing the ill-prepared U.S. armed forces. He motioned for Frank Knox to bring the papers to him. He read quickly through the 12-page document. "What do you think of this deal in its final form, Frank?"

"A day late and a dollar short, Mister President."

"Then, you think I should not sign it?"

"I recommend that you do sign it, Mister President.  I just hope I am wrong, and it is not too late for the British."

"And, how about you, Henry?  How do you view this *quid pro quo* with the British?"

"I agree with Frank, Mister President.  While we must admire the tenacity of the British and especially the RAF, their prospects remain bleak.  However, they are our best hope for time.  If 50, old destroyers, some airplanes and rifles will bolster their defense enough to hold the line, we may just find enough time to be better prepared."

Roosevelt studied Stimson's eyes.  Neither man blinked or looked away.  Stimson's strength and confidence were the very qualities that Roosevelt found attractive and admirable in his Secretary of War.  "You have always been candid, honest and frank with your comments, advice and counsel.  How will Winston view this?"

"Probably about the same as me, although he'll find the bright side."

"You have never tried to camouflage your affinity for the British, Henry.  I share your general views.  It is the public perception of American blood, sweat and tears, or at least no blood yet, supporting and sustaining the British Empire that I find the most troublesome in this business."

"We had seven American citizens in the RAF, Mister President.  If our information is correct, two, maybe three or four, have been killed already."

"You know what I mean, Henry," Roosevelt said with some irritation.  "We are doing everything we can, given the weight of American public opinion and the mood of Congress.  I am up for reelection this November, let's not forget."

"I am reminded everyday, Mister President.  Some principles are higher than self."

"While I do appreciate your candor, sometimes your bluntness does come across as rather abrasive."

"No offense, Mister President."

"None taken.  The only other choice is, Wendell Wilkie, the darling child of the . . . ," he hesitated, wanting to remonstrate his opponent's political party, but decided to avoid offending the political affliation of his two principal war lieutenants even though they did not agree with Wilkie either, ". . . the isolationists.  Even though I may be slow and overly cautious, I am the best supporter Winston has at the moment and may ever have."

"Yes, sir."

"I truly admire his courage, his gumption, his balls, if I may speak plainly."

Both lieutenants laughed.  Roosevelt smiled.

"Have you been able to read the transcript from Ed Murrow's most recent broadcast?" asked Frank Knox, probably wanting to distance the conversation from Roosevelt's partisan political thoughts.

"Yes, I did.  As I recall that boy's name . . . Drummond was it? . . . on the few pardon letters I signed several weeks back."

"Yes, sir.  You are correct," responded Stimson.  "Brian Drummond is his name, from Wichita, Kansas.  The King, himself, awarded him the Distinguished Flying Cross as an ace.  The way Murrow tells it, this kid is one step short of sainthood."

"And, he is probably not even Roman Catholic."

The two men laughed again, but neither lost their seriousness.

"Murrow's commentary and reports are having a demonstrable effect on public opinion, Mister President," added Knox.

"In his inimitable way, he has touched the hearts of our citizens.  We always did have an attraction to the underdog.  Those reports are so grim and stark, and yet vivid.  This young pilot, Drummond, has touched quite a few Americans as well."

"He left for England just after he graduated from high school last year," said Stimson.  "He's been flying fighters for the RAF from the start."

Roosevelt reached for a note pad and began jotting down several items.  "So young," the President mused.  "I'll write a quick note to the lad, his parents, Art Capper and Clyde Reed, their senators, and who's their representative?"

"John Houston, I believe, Mister President," Knox answered.

"Confirm the names for me, Henry, and get me the addresses and a few personal items, if you will be so kind."

"Certainly, Mister President."

"I think they're going to beat the bastards."  Stimson and Knox looked at the President and captured his eyes.  The American leader understood in an instant what the respected New York Republican politician was going to say.  He held up his hand.  "I know, Henry.  I also read General Strong's most recent report.  They are on the ragged edge."  He paused for a moment's contemplation.  "If I thought I could get away with it, I'd have sent as many fighter squadrons as we could spare.  But, if we are to be any help to the British, we must keep that damn ostrich Wilkie out of the White House.  We are doing all we can do."  Roosevelt waited for a response, but both men remained frozen just staring at the President. "The one thing I won't accept is supporting the empire of the British. Whatever support we can muster up for the British simply cannot be seen as sustaining their colonial empire"

"We certainly do agree on that point, Mister President," Stimson responded.

"I don't think the Prime Minister is asking you to," added Knox.

"OK. I've had enough. Please find those names and addresses." He signed the orders that lent 50, old, surplus, World War I, four-stacker destroyers to the Royal Navy in exchange for 99-year leases to numerous naval bases mostly in the Caribbean Sea. "Here you go," he said, as he handed the Secretary of the Navy the papers. "Now, let's get reelected so we can really help Winston. I suspect we don't have much time left before the Germans or Japanese find us too ripe to pass up."

"Yes, sir."

"Henry, I'd suggest you contact Bill Stephenson. He can deliver the papers personally."

"As you wish, Mister President. Anything else?"

"No. That should do."

"Then, if you will excuse us," said Stimson.

The President nodded his head and returned to the papers on his desk. The civilian leaders of the Army and Navy left the President, closing the door behind them.

Alone in the Oval Office once more, Franklin Roosevelt allowed his thoughts to turn eastward. Strong's long memorandum report left little to the imagination. By their own estimates, the RAF was within days of exhaustion or depleted effectiveness. Henry Stimson was precisely correct as he usually was, no matter how painful the truth might be. Roosevelt found himself saying a little prayer for the deliverance of Churchill and the British. If the Germans managed to take Great Britain, the prospects of war with the Germans became enormously greater. As most Americans felt, only the words and defiance from the British Prime Minister, half English and half American, could hold the precarious line of defense against the undefeated juggernaut of the German armed forces. They did indeed need another miracle, after the previous one at Dunkirk.

———

# Chapter 12

Humankind cannot bear very much reality.

-- T.S. Eliot

*Wednesday, 4.September.1940*
*Headquarters, Secret Intelligence Service*
*No.21 Queen Anne's Gate*
*Westminster, London, England*

### Week 9

The conservative, grey suit made Colonel 'C' Menzies appear like a myriad of other businessmen and civil servants in the city trying to carry on despite the rigors of war surrounding them. Like most of the leadership of MI6, and even some of the leaders of other intelligence agencies and government ministers, who chose to visit the Secret Intelligence Service, Menzies used the back entrance to the headquarters building complex. The front entrance and secure lobby stood at No.54 Broadway, across the street from the St. James Park Underground Station. An ordinary, polished, brass plaque marked the front entrance and identified the building as the Minimax Fire Extinguisher Company.

The Queen Anne entrance was maintained in appearance like the myriad of other townhouses in the area and was listed in the postal directory as the office of Coderoy George & Company, Chartered Surveyors, while the official designation was The Passport Control Office. Very few people knew the Queen Anne building served as 'C's residence when he was in London. The objective remained to avoid drawing attention to the purpose of the Broadway House. The high pedestrian and vehicular traffic around the Underground station made detection of surveillance activities far more difficult.

'C' greeted the security personnel, bypassed the upper floor residence, and made his way through the maze of corridors to the Broadway House, fourth floor, and modest office of the director general, overlooking St. James Park. He returned from his Foreign Office appointment in Whitehall, prior to his next scheduled meeting with the Head of SIS Operations, Carl Ambrose Acton, who appeared almost immediately and without announcement.

"I have excerpts from Hitler's speech last night," Acton announced before he reached 'C's desk or took a seat.

For 'C,' reading speeches of the German dictator did not particularly appeal to him. The taut military situation demanded the attention of every intelligence asset to every source of potential information no matter how trivial it might seem. Even though he was the Director-General of the Secret Intelligence Service, Colonel Stewart 'C' Menzies recognized the importance of each piece of information in the greater puzzle.  He was also a political analyst by default.

"Let's have it, then shall we."

Acton began, "Most of the speech to the *Reichstag* contained drivel for local consumption.  However, there were a few words specifically meant for us, and I quote, 'The British drop their bombs indiscriminately and with-out plan on civilian residential quarters and farms and villages.  For three months, I did not reply because I believed that they would stop, but in this Mister Churchill saw a sign of our weakness.  The British will know that we are now giving our answer night after night.  We shall stop the handiwork of these night pilots.'  End quote.  He goes on later in the speech with a passage that we believe is more significant.  Again, I quote, 'Now you will understand that we shall have to give them our reply night after night and in ever increasing strength; if the English declare that they will make mass attacks on our cities, we shall erase theirs.'  End quote."

Acton looked into the contemplative eyes of his Chief.  Menzies' intellect churned through the words along with a plethora of other, often, abstract bits of information.  'C' wanted all this saber rattling to go away, but he knew the saber was now drawn and would soon be used in a rather large way.

"Your assessment?" he asked.

"We believe this may signal a major shift in focus . . . in tactics."

"To the cities?"

"Yes."

"That would not be consistent with an invasion attempt, Carl."

"Correct."

"Then, pray tell, what are all those barges doing in every little harbor, port or fishing village along the Channel coast?"

"He may have given up any notion of invasion this summer.  The season is late after all."

"Do you have any other indicators?  Anything from ULTRA?"

"No.  I did call Alastair this morning."  Alastair was Commander Alastair Ignatius Denniston, Royal Navy -- head of the Government Code and Cypher School (AKA GC&CS, or in more social situations, the Golf, Cheese and Chess Society) at Bletchley Park.  "He has nothing of substance either.  But, we did receive a warning from 'A' Five Four."

"The German agent in Paris?"

"Except he was reposted back to the Foreign Ministry in Berlin recently.  It was rather cryptic.  We have been a bit suspicious of 'A' Five Four lately.  His reposting to Berlin comes after a couple of messages indicating he thought the Gestapo might be on to him."

"What was the warning?"

"He believes the invasion is real, and the landing beaches are in the South between Brighton and Ramsgate followed by a subsequent landing at Dover once the defenses have been neutralized."

"What if he has been compromised and the Gestapo or Heydrich's SD are using him to feed false information to us?"

"Entirely possible, sir. We have no choice given recent reports from 'A' Five Four, but to assume he has been turned or supplanted."

Menzies shook his head in disbelief, as he muddled through the thin and disparate information. "Carl, yesterday, the Prime Minister informed me of the precarious state of our air defenses. The bloody Germans are within days of winning the Battle of Britain, or at least the air supremacy portion of the battle, at any rate. They must know how thin our fighter defenses are. They have but to administer the *coup de grâce*, for God's sake. Why would they stop a winning tactic?"

"Maybe they don't know?"

"What?" Menzies said, nearly to the edge of protestation rather than question.

"Their air force intelligence section is headed by a major, and an inexperienced major at that. Maybe Hitler sees or senses the weakness in his own intelligence apparatus."

"Doubtful but plausible."

"Yes, sir. The truth should come within the next few days given past performance."

"And, you think he will shift his aim from the air defense system to the population?"

"That is what the words say, and a change from daylight raids to night bombing, where they have a distinct advantage at present."

"Dear God above, I hope you are correct, Carl. The Prime Minister, the Air Ministry, Fighter Command and probably every young pilot we have, are praying for deliverance. It is that bad. The pilots, what few we have remaining, are near the human limits of fatigue and exhaustion. We now have new aircraft available, but no pilots to fly them. The Germans have ground down Fighter Command methodically. Everyone believes they are within days of the final condition for invasion."

"The Royal Navy will have something to say," said Acton, more as a reflex than a considered thought.

"Not without air superiority."

"I take your point, 'C.' Then, we wait."

"It would appear so."

"Do you want the text of the full speech?"

"No."

"Will there be anything else, sir?"

Colonel Stewart Menzies rubbed his chin and pulled at his mustache as he thought about the question. "Has 'Jumper' or anyone else heard from 'Diamond' lately?"

"'Diamond' was transferred to SOE a month ago, according to SOE, he is still in Amsterdam. His last report was nearly a fortnight ago."

"Why don't you take a walk over to SOE and share your assessment of Hitler's speech? Talk to Dalton or Gubbins, maybe 'Diamond' can give us a view of activity in the harbor area. We need some elements of confirmatory intelligence."

"Maybe it will be the bombing of London?"

"Only if it were in a big way, such as abandonment of the air defense assault."

"I shall see what I can do over at SOE and I will check with Admiralty as well to see what Admiral Pike might be able to help with on this matter."

"Excellent. We should sit on this until we have more evidence."

"As you wish, sir."

———

*Thursday, 5.September.1940*
*RAF Middle Wallop*
*Middle Wallop, Hampshire, England*

"I just got up. Why am I so bloody tired?" Brian asked his best friend, Jonathan Kensington, as they plopped down in two lawn chairs in front of the dispersal tent.

"It's called too many bullets and not enough sleep, you twit."

"No need to get hostile."

Jonathan chuckled. "That was a rhetorical question, correct?"

"What's that?"

The slightly older British pilot started to say something, probably derogatory, and then stopped. "A question that isn't a question."

Brian thought about the words for a moment. "Then, you're right, it was a rhetorical question."

"See, I am so tired I can't tell."

"How much longer can this go on?"

"Not much, I should think.  That poor bloke in Two Three Eight Squadron landing with his undercarriage still in the wings, and several others nearly falling victim as well.  We are all so bloody exhausted, we can't think straight anymore."

"And now, we've got airplanes, brand new ones, sitting here," Brian said, nodding toward the flight line, "with no pilots.  We don't have a replacement for poor old 'Slim' Koenig, and we're probably not going to get 'Curly' Mansek back anytime soon."

"Why don't you bleeding bastards find something else to talk about?" said Flight Lieutenant John 'Waggle' Davies.  "If you keep at it, we shall all be ready for the asylum."

Brian and Jonathan looked at each other, smiled and shrugged their shoulders, as if to say, 'what's his problem?'  The normal frivolity, devil-may-care, thumb-your-nose-at-danger, attitude of the squadron pilots disappeared with the mounting fatigue and lack of relief.  Joints ached from the cold of high altitude flight.  All the pilots seemed to have a perpetual headache.  They all longed for poor weather and the shortened days of winter.  Everyone asked themselves, if they would make it to winter, although none of them would talk about their feelings, not even Jonathan and Brian.  To the young American, it was like a chronic illness – talking about it did not make it better or make it go away.  Brian also felt shades of guilt at moments when he could recall with an internal smile the unintended rest break he enjoyed on Monday.  Most of his comrades had not been able to take a 24-hour break in several weeks.  Brian had been the last pilot to have a break and probably would remain the last for a while unless someone was injured.

The telephone rang twice before Corporal Jennifer Warren lifted the handset.  Brian found himself instinctively looking to the corner of the tent for 'Slim' Koenig to be hunched over vomiting his breakfast, and then felt bad for the loss of his fellow American.

"Scramble, scramble, scramble," shouted Warren.

The groans and protests punctuated the lethargic motion of the pilots.

"Snap to it, lads," shouted Squadron Leader Darling with a streak of anger in his voice.  "Gerry calls us to a party.  We shant disappoint him."

The No.609 Squadron fighters joined the air battle southeast of Southampton.  Bombers and fighters were scattered about like a bowl of mixed nuts.  Airplanes maneuvered through many thousands of feet of altitude.  The Middle Wallop pilots waded into the confused engagement, each trying to find a target.  A significant danger in these crazy brawls came from friendly guns.  Every pilot found himself more than once resting his gunsight reticle on a sleek monoplane that did not have the correct shape, and then quickly pulling off looking for a proper target. The experience of Fighter Command dropped as young, inexperienced, scared pilots were thrown into the desperate air battles over the English Channel and Southern regions of England. Brian had seen the business end of several Spitfires and Hurricanes over the last few weeks, requiring him to maneuver hard to avoid any mistake. Mid-air collisions took their toll more frequently, as the numbers of aircraft in the same block of sky increased to unimaginable levels.

Both sides fed more aircraft into the large battle. The radio communications became just as confused as the maneuvering aircraft. Words flew through the air like bullets sprayed across the sky – desperate fragments from a mortally wounded man screaming toward the ground – calm words of encouragement – warnings of unseen danger. The radio became useless as a means of communication. Brian found himself ignoring and blanking out the bewildering jumble in his ears.

The battle continued through most of the morning with a strange ebb and flow of fighters. No.609 Squadron expended their ammunition and nearly all their fuel before they diverted to RAF Tangmere, now a heavily bomb damaged airfield barely remaining operational. Fighters crowded the available space waiting for refueling and rearming. Most of the pilots, including Pilot Officer Brian Drummond, stayed strapped into their seats taking the moments lull in combat to catch whatever sleep they could. The ground crews, who did not know most of the aircraft or pilots, struggled against their own fatigue to return the fighters to full capability as quickly as possible without bothering the dozing pilots.

Brian felt a hand shaking his shoulder. Startled, he looked up through his goggles still over his eyes, into the face of a concerned crewman. "Sir, your squadron is cranking."

Brian nodded his head and mindlessly stepped through each item required to bring life back to the powerful Merlin III engine in front of him. He found 'Jackstay's 'PR-D' Spitfire and 'Crazy's 'PR-S' fighter, and took his position for takeoff. Instead of a squadron takeoff as they commonly did over the last few weeks, they took to the air in sections one behind the other using the only reasonably smooth patch of ground available. Within minutes, they were back into the same block of airspace they recently departed. As the squadron maneuvered to join the fight, Brian could not see any bombers in the clear skies. The engagement had become all fighters from both sides.

As the reality of the situation sank in, the British fighters were directed by the controllers to disengage and withdraw.  The command produced an instant of resentment that passed quickly with his recall of the many conversations about the German intentions to incapacitate the British fighter defenses. Bombers did the damage on the ground.  Fighters depleted the thin protective cover.  The Germans had insufficient fuel to give chase unless they chose to become guests of the British for the duration of the war.

The lull allowed them to return to RAF Middle Wallop, after the lunch hour, again.  The irritation of no food had no discernible effect other than Brian's growling stomach.  To the pilots, sleep and some attempt at recovery became the dominant need.  Lack of food would not kill them in the air; fatigue induced lack of concentration and focus would certainly lead to their demise.  The margin of error usually allowed for mistakes or lapses in attention dropped to zero, weeks ago.

One of the junior officers from the station operations section drove to their dispersal area in a dark green Land Rover.  He carried a small folder to Squadron Leader Darling who struggled to sufficient alertness to receive the folder from the pilot officer.  Brian watched him read the contents with indifference.

"Listen up, lads.  Group wants us to see this message from One One Group," he said.

Grunts and minor protestations provided no clue as to the content until Brian took his turn to read the message.

---

## SECRET

```
ZZZZ/5418ATB3211/FC-11/050917/TTGPPBWQ/573/ZZZZ
DATE:   05.09.40, 1037 HOURS
FROM:  HQFC
TO:  ALL FC GROUPS, STATIONS AND SQUADRONS
RESEND
DATE:   05.09.40, 0900 HOURS
FROM:  HQ, 11 GROUP
TO:  11 GROUP
SUBJECT:  PROTECTION OF AIRCRAFT FACTORIES
BREAK
ENEMY ACTIVITY FOCUSED ON DESTRUCTION OF ADGB
PROBABLY IN PREPARATION FOR INVASION ATTEMPT.
MAINTENANCE OF AIR DEFENCE SYSTEM VITAL TO
```

```
NATIONAL DEFENCE.   PROTECTION OF AIRCRAFT
PRODUCTION FACILITIES CRITICAL.   UNTIL FURTHER
NOTICE, MAXIMUM EFFORT SHALL BE MADE TO PROTECT
AIRCRAFT MANUFACTURING WORKS WITH PRIORITY TO
WOOLSTON, BROOKLANDS, AND HATFIELD.   EXPECT
DIVERSIONS TO FULFILL THIS OBJECTIVE.
END
ZZZZ/5418ATB3211/FC-11/050917/TTGPPBWQ/573/ZZZZ
```

**SECRET**

---

Brian recognized the Vickers-Supermarine Spitfire Works at Woolston near Southampton and the deHavilland Works at Hatfield, as well as the Hawker Works at Brooklands, which was still the sole manufacturing facility for the Hurricane. What about the many other aircraft factories? Why hadn't they identified Bristol, Yeovil, Derby and all the others needing protection? He wondered what else was happening to prompt such a message, and why would they want the pilots to know their tactical objectives? The pilots felt for sometime they were simply instruments like bullets.

"What do the friggin' blokes think we've been doing?" asked Flight Lieutenant John 'Waggle' Davies in a sour tone of a tired man disturbed from his sleep without cause.

"Not enough," answered Flight Lieutenant Roger 'Jackstay' Beamish.

The brief respite did not last long. They returned to the air for another large brawl South of Southampton. No.609 Squadron continued in the fight until fuel exhaustion and prudence dictated disengagement. While the battle managed to keep the Germans from their intended targets, Brian knew it was more to hold on for the draw than try for the knockout. The seemingly endless grind dulled the senses, softened the reflexes and dissipated the fight in each of them. They were down to simple survival, now. Fortunately, the day's encounters came through to a draw – no one was shot down or injured, but there were no victories either.

As they usually did, the No.609 Squadron pilots along with most of the aircrews in Fighter Command tried to wash away the pain in their heads, joints and muscles with alcohol and humor. Everything seemed to be wearing thin. Brian and Jonathan wanted the soothing touch of the women close to them, but neither of them could muster up the strength to do more than lift their beer glasses until they were empty. The endless tunnel remained dark and foreboding.

—

*Friday, 6.September.1940*
*SS Duchess of Richmond*
*Pier 23*
*Halifax, Nova Scotia, Canada*
*07:30 hours*

The Atlantic Ocean crossing had been without incident, although observing their destroyer escorts dart around the convoy a few times had been disconcerting. The running commentary of the Royal Navy sailors who happened to be along the railings at the time did not lessen the anxiety of being aboard a ship that was clearly a target for any lurking German U-boats. The seas had been comparatively calm, as the sailors pronounced, although the deck remained in constant motion and produced more than a few cases of seasickness among the scientists and engineers of the British Technical and Scientific Mission to the United States of America team, now also known as the Tizard Mission, named for their leader Sir Henry Tizard, who had been in Washington, DC, for the last few weeks, completing the final coordination with the Americans. As they passed McNabs Island and entered Halifax harbor, the prevalent Atlantic swells had disappeared.

The Tizard Mission team had taken to the Observation Deck railings to see historic Halifax harbor, as sailors for decades had witnessed. The sailor's city had grown on the port side. They could make out the many busy piers that made the harbor so valuable. They could also see the comparatively new buildings further up-harbor that replaced the massive destruction of the Pier 6 explosion of the SS *Mont Blanc* at 09:04, on Thursday, 6.December.1917, after a collision with another cargo ship – SS *Imo*. The fully loaded *Mont Blanc* contained six million pounds of high explosive ammunition produced in Canada and destined for the battlefields of France. The ship became one big bomb – the largest man-made explosion in history. Beyond Halifax city, a couple of the knowledgeable naval officers had pointed out the entrance to Bedford Basin, where convoys gathered in safety before transiting the harbor for their perilous journey across the Atlantic Ocean.

As the *Richmond* had approached Georges Island just after dawn, she turned to port, picked up two tugs and the harbor pilot to slowly dock bow first at Pier 23, one of the primary railroad terminal piers servicing Halifax harbor. Once the ship was fully moored, the deck crew began the process of off-loading their passengers and cargo. The team was not scheduled to disembark until nearly noon, when they would leave the ship, as their precious crates, baggage and lockers were loaded and secured on a special combination passenger and freight train at the dock.

Royal Navy sailors began to appear on deck at the railings with them. A navy lieutenant informed the team members the first of eight of fifty U.S. Navy destroyers to be 'lent' to the Royal Navy in exchange for long-term lease rights to British military facilities, mostly in the Western Hemisphere, were due to tie up just across the corner at Pier 22 from them, to facilitate the inspection and transfer process. As they waited, they listened to stories of the convoy escort experiences the naval officers had since the war began a year ago. The civilians were thankful they had not heard the stories in Liverpool, before they made the crossing. Watching the bombing of Liverpool on the night of their departure was bad enough. What the naval officers experience every few weeks seemed far more threatening. There was nowhere to hide in the middle of an ocean.

They did not have to wait long. The morning continued to brighten. The first, distinctive, gray silhouette of a slim, small warship came into view, moving slowly into the harbor. The four, boiler, exhaust stacks of the flush-deck, Wickes-class, Great War era destroyer displayed a large, white, hull number 198.

The Royal Navy commander assigned as the squadron commander for the flotilla, referred to a single piece of paper. "That is the *Herndon*," he said. "She will be re-designated the *Churchill* when the transfer is completed."

One by one, additional similar destroyers appeared in trail behind the *Herndon*. All eight warships made an impressive image. The same two tugs that docked the *Richmond* began to the same for the *Herndon*. The docking sequence took a little over an hour in a well-choreographed routine. The *Herndon* moored to Pier 22. The other seven destroyers moored side-to-side to the *Herndon*. The stack of those warships meant more than their physical presence.

Their attention on the soon-to-be additions to the Royal Navy did not last long. They all recognized the large crate containing a fully functional and equipped Rolls Royce Merlin XX engine – the latest model of the respected, liquid-cooled, aircraft engine. The 1,480-shaft horsepower motor was not yet in operational fighters, but was installed in numerous, prototype aircraft like the Vickers-Supermarine Spitfire Mark III, still undergoing qualification testing. The other crates and lockers that comprised their cargo followed. They watched a dozen plus Royal Navy officers walk across the wharf to be greeted by U.S. Navy officers, as the process of exchange began.

"It is time for us to disembark and board our assigned railcar," announced Royal Navy Captain Hugh Faulkner, their transit leader.

The team abandoned their railing perch, descended below decks to vacate their staterooms, and made their way to the passenger railcar, one of two, along with a dining car, and followed by two, flatbed railcars.  The last of their cargo was loaded on the railcars under the watchful eye of a dozen, armed, British, Canadian and American soldiers.  The stevedores covered both cargo railcars with heavy canvas, tied down tight and then covered with heavy security straps.  Shortly after noon, the train began to slowly move through the port's railway, marshalling yards for their journey south into the United States and to Union Station, Washington, District of Columbia.  So began the next phase of their historic mission.

—

*Friday, 6.September.1940*
*Headquarters, Secret Intelligence Service*
*54 Broadway*
*Westminster, London, England*
*11:30 hours*

The Joint Intelligence Committee gathered the various British intelligence agencies to ensure the best possible analysis was provided to His Majesty's Government.  Deputies handled the daily operations of the committee to ensure the maximum integration of intelligence collection and analysis.  On this particular day, time and circumstance, the chiefs agreed to meet, not unusual but not routine either.  The situation warranted the gathering.

In the secure conference room, large-scale maps of the British Isles, the North Sea, the English Channel, Western and Eastern Europe, the Mediterrean Sea, North Africa, the Middle East and the Atlantic Ocean covered all four walls of the windowless room.  Seated at the main, retangular table were Director General MI6 Colonel Stewart Menzies, 'C' to his colleagues; Vice Admiral Sir Geoffrey 'Jumper' Pike, Director of Naval Intelligence; Major Gary Templeton, chief of the Army Intelligence Corps; Air Vice Marshal Courtney Medford, Head of the Air Intelligence Department, and Brigadier 'Jasper' Harker, Acting Director General MI5 (Security Service).  Head of GC&CS Commander Alastair Denniston had been specifically invited to join the meeting, and MI6 Director of Operations Carl Acton would conduct the foundation briefing.

"Gentlemen," Menzies said, "let us get this meeting started.  Before I turn the floor over to Carl, I must convey my appreciation to each of you for taking the time out of your busy schedules to attend.  I met with the Prime Minister yesterday and the War Cabinet the day before yesterday.  I think it fair to say, the government remains quite apprehensive regarding the rapidly escalating signs we are seeing across the Channel.  The War Cabinet is expecting our assessment and recommendations this afternoon."

"Well, then, let's hop to it," Pike interjected.

Menzies nodded his head to Acton.  Carl stood and moved to the head of the table opposite 'C' and in front of the screen upon which he intended to project several aerial photographic transparencies produced at unreasonable cost for this meeting.

"We have a rapidly changing, complex situation in the Channel operating area.  The Germans continue to carry out consolidation operations in Northern and Western France intended to secure their ill-gotten gains.  Those operations appear to be tapering off based on a wide variety of reports from our remaining few agents in France and friendly sources, who are able to communicate with us by one means or another.  Troop movements, mostly by rail, some by lorry, all point to assembly areas inland from and to either side of the Pas de Calais.

"The first image I have to show you," Acton  said, nodding to the projectionist, "was taken yesterday over Boulogne-sur-Mer."  The black and white aerial photographic image illuminated the room as the projectionist switched off the overhead lights.  Acton had retrieved a yard long wooden pointer.  "This image is illustrative of several dozen photographs taken by reconnaissance aircraft over the span of the last week.  In this one image alone," he said, pointing to each area, "we see by our count 49 various size river boats gathered in the port, despite the on-going, rather feverish, engineering efforts to clear the battle damage from last spring and provide new, large-scale, logistics facilities.  We also see here and here," he said and pointed two areas perhaps a mile inland, "two divisional bivouac sites, one camp newly installed and the other partially camouflaged."  He pointed to a repaired and clearly operational airfield.  Two Bf-109 fighters could be seen in the image as they were taking off.  "The main aerodrome at Boulogne was returned to full operational status before Eagle Day last month.  The base now has an entire wing, six squadrons, of fighters at this aerodrome alone.  They have even built several revetments and are building more, as well as new petrol storage tanks, which would suggest they plan to conduct vigorous combat operations."

"I presume from your choice of words," interjected Pike, "you have similar images from other sites."

"Yes, sir . . . six sites from Dieppe to Ostend," Acton responded. "Even within that span of coastline, we have not yet covered each potential embarkation site, due to intervening weather."

"But, you have reason to believe we will see similar build-up of forces," Pike added.

"Yes, sir," answered Acton.  "Limited ground observations from the few friendly sources we still have in the area suggest the gathering of forces for amphibious operations is being carried out at least from Le Havre to Antwerp . . . and perhaps even beyond that range.  We also have several observations of what can best be described as embarkation and landing training.  The German ground forces located along the coast appear to be conducting general training for amphibious operations."

"We have been working on refining the German order of battle," Templeton began, "based on the latest information from all known sources."

"Are there unknown sources you are deriving information from?" asked Menzies with a smile for the youngest among them.

"Point taken, sir . . . information from all sources," Templeton corrected his statement.  "The coastal assembly areas, as Mister Acton displayed, suggest 20-22 infantry and armor divisions have been allocated to the initial phases of Operation SEALION, leaving a handful in reserve and for defense; they are all first line, combat experienced units."

"And, their best field generals," Menzies interjected.

"Yes, sir, that as well," said Templeton.  "I must add . . . the Germans have stockpiled what we believe to be 30 days of combat supplies, and they continue to build their supply levels."  An odd silence filled the room.

"And, three, full, reinforced, air fleets," Medford mumbled.  They all looked to the only RAF officer in the room.  The seven intelligence service leaders actually laughed, as if it was some inside joke realized.  They allowed the laughter to persist until it tapered off naturally.

"We . . . ," began 'C,' " . . . we have certainly borne witness to the magnitude and destructive energy of those three air fleets, have we not?"

"And, they are far from done with their dastardly deeds," Medford added.  "The most recent status report and estimate from Dowding and Fighter Command from two days ago reports that the air defense system of Great Britain is within days of collapse.  The fighter force has been operating without reserve of any form for several weeks now.  They have cancelled all leaves, rest, rotations and transfers.  They have moved every available pilot into front line squadrons, and they are continuing to be ground down and literally bled to death.  If something does not change and very soon, the fighter force will be unable to carry on defensive operations and even less capable of conducting offensive operations against an invasion fleet."

"Where does that leave us?" 'C' asked.

Carl Acton picked up the gauntlet.  "Our analysis of the probable German conditions necessary for their cross-Channel invasion decision rests upon their assessment of conditions.

"The first prerequisite is likely the assembled forces and materiel for an amphibious operation."

Denniston interjected, "Various sources suggest the German High Command views a cross-Channel operation as nothing more than a river crossing."

"A twenty mile wide river," interjected Templeton. They all laughed, again.

"Quite so," Acton said. "Perhaps the Germans underestimate the complexity and difficulty of crossing the English Channel . . . which offers us some modicum of advantage. Nonetheless, I think we can all agree they have completed that portion. The estimates of ground forces assembled along the Channel coast are sufficient for a successful landing. Given the condition of the Home Forces after the debacle in Northern France, there are little of our conventional forces to stop them, if they successfully establish a beachhead, and they are likely over-estimating their advantage. We really need to see what is behind the invasion force to sustain their landing force, should they decide to execute SEALION.

"The next prerequisite is air superiority. I think we can agree, they are nearly to that point, and they may well have concluded they have attained sufficient domination over the southeast of England to support a cross-Channel operation. The conservative assessment would be to assume they have decided that condition has been met at least over the likely landing beaches.

"The only prerequisite left is naval superiority," Acton said.

"This is where I come in," Denniston announced. Not everyone in the conference room was cleared for and had access to the ULTRA source material. "Our sources indicate the Germans remain apprehensive about the Navy's ability to protect the invasion fleet. The First Sea Lord has deftly deployed the Home Fleet to reinforce the uncertainty, as reflected in the German apprehension."

"So, where does that leave us regarding the naval prerequisite?" asked 'C.'

Admiral Pike picked up the argument. "The Admiralty agrees with Commander Denniston's representation. The First Sea Lord has alerted the Home Fleet to be prepared to immediately deploy two cruiser squadrons, a heavy flotilla of one battleship and two battlecruisers into the North Sea within hours of our execution signal. We are also keeping a very close eye on the North Sea German ports for unusual deployments. So far, we have no evidence the Germans have sortied for general action. Based on that, we do not see indications of immediate action, in other words, not within the next six to twelve hours."

"We've not heard from you, 'Jasper,'" Menzies said.

Harker looked around the room. He was also not aware of everyone having access to ULTRA. "Reliable sources indicate the German security services may already have comprehensive post-invasion plans to dominate this country." He wanted to discuss the details of the ULTRA message decoded three days ago, but he could not. "The Security Service has noted an increased quantity of infiltration attempts in the last month, which indicates to us that something significant is afoot. We continue to interrogate recently captured agents, and captured pilots and aircrew members. The image that is beginning to emerge from our interrogation efforts indicates surprise that the invasion has not already been carried out. I must report a particularly consistent tone of arrogance and dominance among the Germans who are talking. They apparently feel little need to protect knowledge of pending operations. Several of the pilots actually believe we are withholding information from them regarding the successful progress of the invasion by the heroic armed forces of Germany."

"You have to love the audacity of the Germans," Pike commented.

"They apparently believe the tripe the National Socialists have fed them for so many years . . . Aryan superiority and all," Medford added.

"Does anyone have anything else to add?" Menzies asked.

"There are mountains of information of similar content," answered Pike. He looked around the table. "I suspect all of us see the information we have in the same light."

"Carl?"

"I agree with Admiral Pike," Acton responded. "I have additional aerial photographic evidence as well as other field sources that are essentially the same."

"Can or should we draw any conclusions from what we see today?" 'C' asked. "And, what are we to recommend to the War Cabinet this afternoon?"

Admiral Pike stood, and paced back and forth several times before turning to face the group of intelligence chiefs. "The signs of impending invasion. All of the information we have suggests we are days, not hours, from an attempt by the Germans to land substantial ground forces in Southeast England. We do not see signs of immediate invasion activity, such as shore bombardment by large caliber warships and loading of their transport craft. I doubt the Germans would be assembling these forces and equipment as some sort of demonstration or attempt to intimidate us. Their efforts have only one purpose – the invasion of England. The War Cabinet should know the signs we see. As far as a recommendation, it is my opinion that the War Cabinet should issue the appropriate order to bring the Home Forces and Home Fleet to maximum readiness and alert. We may have mere hours to react once they decide to embark their invasion force and concentrate their warships to support amphibious operations."

"Very well. Courtney?"

"I agree completely with 'Jumper,'" Medford answered. "I would also suggest we add a notation regarding the aerial bombardment of the enemy's assembly areas. Such large forces and staged equipment are ripe and inviting targets for aerial bombardment. While I think we all recognize anything beyond our observations and analysis enters the domain of operations, I am quite concerned that desperate times will lead desperate men to do desperate things. The fighter force is fully engaged, and it can be argued that they are no longer able to defend the skies above this country. The priority must be air defense, which in turn means any bombers deployed to attack the enemy assembly areas would be unescorted – a very costly endeavor. I remember vividly the tragedy of our efforts to attack the River Meuse bridges during the early days of the Battle of France, last spring."

"Good observations, Courtney, and I suspect we all agree with your assessment, but we must remained focused on the facts as we have them. The leadership will have to make the operational decisions. If we are agreed, I will report our concerns."

"I can support that," Admiral Pike added.

"Army intelligence can as well," Major Templeton said.

"Alastair, Carl, your opinion or position?"

"I have no reason to disagree," responded Denniston.

"Me, either," Acton contributed.

"Very well. I shall report our assessment as unanimous. Does anyone wish to add recommendations to assist the War Cabinet in taking appropriate action to convince the Germans not to execute their Operation SEALION?"

"Personally, I think Sir Hugh Dowding pegged our counsel precisely last spring," Pike said. "We are too thin to take offensive action. We must husband our precious resources to repel the invasion that appears nearly inevitable at this stage. It is only a matter of when, I suspect. If we do nothing, we must impress upon the War Cabinet the immediacy of what we see. The Germans have perhaps another few weeks of decent weather to make their attempt. The Spanish did not fare so well with the weather in 1588, and we are past that time of year already. The trial we have endured for the past several months is drawing toward the conclusion. I suspect we shall have the verdict in the next few weeks. Our margins are extraordinarily narrow. The Prime Minister, the War Cabinet, our ministries and indeed the entirety of His Majesty's Government deserve the best, most accurate information we can possibly provide them to sustain us through to the conclusion of this trial."

"Well said, Sir Geoffrey," Medford offered. "I think you chose the most appropriate example. Sir Hugh was under enormous, perhaps even incalculable, pressure to send 10 first line fighter squadrons to France during the height of the battle. He calmly convinced the War Cabinet that he was already short of the necessary fighter squadrons for the defense of the Home Islands, and further, sending more precious fighters to France would not alter the outcome. Given our state today, and what 10 less squadrons would likely have done to our air defense situation, Sir Hugh was spot on accurate. I would like to offer my opinion, as well. Intending no disparagement to the Navy, Hitler is least likely to respect Raeder's apprehension about his Navy's ability to fend off the Royal Navy. If Göring, Keitel and Halder say go, he will likely overrule Raeder and launch SEALION. To put this bluntly, our only hope is to put up sufficient air defense capacity to convince Halder, the risk of air attack during the critical phases of an amphibious operation is too great . . . to cause them to hesitate, to delay a cross-Channel attack."

"You yourself told us we are nearly at the end of our viable capacity, Courtney," Menzies said. "How do you propose we do that? What am I to tell the War Cabinet when they ask the obvious question?"

"Keep attacking Berlin. It may well be sufficient insult to Hitler's sense of power that he alters his focus on the air defense system to something else, anything else . . . the landing beaches, the cities, anything."

"Now, that is a particularly dreadful suggestion," Templeton interjected. "You are advocating shifting the enemy's targeting from the Air Force to the very people you are supposed to be protecting."

Medford ignored the jab. "We need time . . . for the fighter force to recover."

"We may not have time," Pike said softly with an aire of resignation.

"We must find the time. Our fighter pilots have kept the Germans at bay, so far. We must find the means to help those pilots."

The room reverted to silence, again. They considered the words exchanged.

"Very well," 'C' said. "Unless there is any disagreement or objection, I shall do my best to convey our assessment, findings and recommendations." Menzies scanned the room; each of them nodded their head in agreement. "With that, I think we are adjourned. Thank you, again, for making time for this meeting today. May God give us sufficient strength to hold the Hun at bay, at least for the next few weeks."

The leaders of the intelligence services stood, gathered their papers, and traded words of social camaraderie and mutual respect. After a few minutes, they began to make their way out of the Broadway complex.  Several chose the front entrance, while the remainder used the back door, Queen Anne's Gate exit, where their automobiles and drivers waited patiently.  Each of them wondered how the next few days and weeks would play out.

—

# Cap Parlier
### Author

—

Cap and his wife, Jeanne, live on the Great Plains of Kansas, along with two dogs and a cat. Their four children have begun their families as well as producing grandchildren – the rewards of long life. Cap is a graduate of the U.S. Naval Academy, a retired Marine aviator, a Vietnam veteran and an experimental test pilot. Cap has retired from the corporate world and can now fully indulge his passion for the story. He has numerous other projects completed, but not yet published, and others in development, including screenplays, historical novels and a couple of history books.

—

Interested readers may wish to visit his website at http://www.Parlier. com for his essays and other items, or subscribe to his weekly Blog: *"Update from the Heartland."* Cap can be reached at: Cap@SaintGaudensPress.com.

SAINT
GAUDENS